C000079188

Pawprints Through my Life

Christine Paradine

Pawprints Through my Life

MEMOIRS

Cirencester

Published by Memoirs

MEMOIRS
PUBLISHING

Memoirs Books

25 Market Place, Cirencester, Gloucestershire, GL7 2NX
info@memoirsbooks.co.uk www.memoirspublishing.com

Copyright ©Christine Paradine, February 2012
Email: chrisparadine@gmail.com

First published in England, February 2012
Book jacket design Ray Lipscombe

ISBN 978-1-908223-97-5

Printed in England

Pawprints Through my Life

CONTENTS

To my grandchildren
Will, Jazzy, Scott and Emily,
with love.

INTRODUCTION

I first became involved with the rescue and care of animals in 1990, because I needed something more in my life and had always loved animals. I joined the team at Animals in Distress on a voluntary basis and a few years later I helped to set up the rescue centre in Ipplepen, near Newton Abbot in Devon. For twenty years, this work enabled me to devote a large part of my life to caring for abandoned and neglected animals of all kinds.

In 2010 my association with the charity came to an unexpectedly abrupt end when through 'management reorganisation', I was made redundant. However, as my family and friends assured me at that difficult time, life goes on. I will always have my memories – happy and sad, good and bad - of the people, events and, especially, of all the wonderful animals I came to know.

This book tells of some of the experiences, challenging, emotional and amusing, from my time involved in animal welfare and of some of the wonderful dogs, cats and other creatures I have encountered.

During my lifetime, I have met many remarkable people, many of whom have become close friends. Just a few of those I thought were friends turned out not to be. I have also encountered many animals - who, without exception, have never betrayed my friendship. Perhaps that is why I have always loved animals!

CHAPTER ONE

A PASSION FOR ANIMALS

Me, 2 years old, meeting a tortoise with Uncle Vic

I have always loved animals. Even as a very young child, I was desperate to have my own dog. Unfortunately, domestic circumstances made it impossible. When I was born, we were living in a ground-floor maisonette in a quiet suburb of London. My Mum, Dad, sister Joan and I shared one small bedroom, and my first bed was a wooden drawer.

My Mum explained to me in later years that it was not because we were poor - it just seemed practical at the time, as a drawer took up less space than a cot. Fair enough.

When we moved to a two-bedroomed bungalow in Kingsbury, north west London, it seemed like a palace to me. However, although we now had more space and our own garden, my dream of having a dog was still not fulfilled, as Mum and Dad both worked long hours and Joan and I were now at school.

A friendly old lady who lived round the corner had a little grey poodle called Johnny, who I thought was the best dog ever in the whole wide world. I used to regularly knock on her door and ask if Johnny could play. Sometimes she would invite me in and I would play with my little friend in the garden for hours on end.

I used to fantasize that Johnny was my dog, and walk down the road holding a pretend lead with my own imaginary Johnny on the end of it. I stopped at lamp posts to let him have a sniff and cock his leg, and made him sit before I crossed the road. He was very well behaved, of course, and came everywhere with me.

When we were having dinner, I would imagine him lying by my feet under the table and, in the evenings I would pretend he was curled up on the rug in front of the coal fire. I can't remember how long my make-believe Johnny was with me, but I certainly enjoyed the pretence.

Duncan, the imaginary horse

Another of my imaginary pets was Duncan, a big brown horse. In his case, he didn't exist at all – he was entirely a product of my imagination. But I can still see him in my mind's eye. When I went shopping with Mum, Duncan came along, and I would pretend to tie him up outside each shop while we went inside. It must have looked very strange to passers-by but my Mum was very kind and patient and let me get on with it.

Duncan even came on holiday with us to Higher Marsh Farm in Dunster, Somerset. I pretended to tie him to the car door handle and imagined him galloping as we drove along. He must have had a good turn of speed to have kept up with us at 70 mph in the outside lane of the A303.

Me and Duncan at Higher Marsh Farm

One day we all went to nearby Minehead for a boat trip. I 'tied' Duncan up to some railings at the harbourside because, of course, a horse couldn't go on a small boat. That would be ridiculous!

We had a lovely cruise around the coast, but on the way back the boatman asked if anyone would like to be dropped off at the other end of the bay. 'Yes please' said my dad. When we got off the boat I burst into tears, because of course Duncan had been left tied up at the harbour. Disaster!

I was inconsolable. Eventually Dad went all the way back to the harbour with me, about half an hour's walk, so we could pretend to collect Duncan. What a lovely Dad. In later years he insisted that it had been my mum who had made that long walk, but I'm sure it was him.

I also used to pretend that my scooter was a horse. I used to gallop around the streets near our home for hours, and if I had to leave the scooter anywhere I would make sure the front wheel was near a patch

3

of grass so my horse could feed. Oh dear, what an imagination!

I have always loved the feeling of fur, so soft and warm. Even now, I can never resist the urge to touch it. Feathers and fish scales just don't have the same appeal.

My Mum told me that when I was very young and we were out shopping, if she couldn't find me, she would look around for a lady nearby wearing a fur coat. I would be there behind her, discreetly stroking and snuggling up to a complete stranger. Just as well that I had no idea the coat must have been made from animal skins.

During my childhood in the 1950s, it was not unusual to see dogs roaming the London streets, sometimes on their own, sometimes in packs. Their owners would let their pooches out in the morning and they would roam the streets all day until it was time to go home for their dinners. The dogs in our neighbourhood were usually quite friendly, and I knew most of them by name.

I'm not sure how I got away with it, but, if I saw one of the local dogs out on his own, I would call him to me and then take him home by the collar. I would then ask the owner if I could take their dog out for a walk on his lead, for which I would charge a shilling an hour (5p today). I would proudly walk around the streets, pretending I was the dog's owner. I had several regular customers, and made enough money each week to buy a good supply of sweets. Thinking about it now, I imagine the poor dogs must have run in the opposite direction when they saw me coming!

Holiday encounters

I have wonderful memories of our family holidays in Dunster, and we went there several times. Joan and I shared a bedroom in the farmhouse attic and a ceramic bowl and jug of hot water were brought up to our room every morning for us to have a wash, as there was no running water.

My sister Joan, me and Dad + sheep with bandaged foot

It was a working farm, so there were lots of animals wandering around the farmyard including cats, rabbits, pigs and chickens. I especially remember feeling very sorry for a huge sheep with a bandaged foot.

At about six o'clock every morning, my Dad would stand in the farmyard under our bedroom window and whistle the song Bali Hai (from South Pacific), our secret signal. If I was awake, which I usually was, I would creep downstairs and join him for a walk before breakfast. We would walk quietly through the enchanting country lanes surrounding the farm, often without talking at all, and see what wildlife we could discover. I can remember the thrill of seeing a family of wild rabbits grazing peacefully in the warm early morning sunshine and once we saw a kingfisher darting along the banks of a stream – the only time I have ever seen one in the wild.

If we had time, we would take the lane which led down to the beach. That was my favourite walk. On the way, we passed a field where a friendly horse lived and we would feed him the sugar lumps that Dad had taken from the sugar bowl during last night's dinner. Dad taught me the correct way to hand-feed a horse - 'hand flat and keep your thumb out of the way!' He knew everything, my dad. I loved the feeling of the horse's soft mouth on the palm of my hand. It was a bit slobbery, but I didn't mind. We usually rambled through the lanes for about an hour, then return in time for a huge farmhouse-cooked breakfast. Delicious!

I thought it was a wonderful, magical place and wanted to stay there forever. To this day, the aroma of a farmyard always brings back memories of those happy family holidays.

Turkeys and tortoises

Another memory from my childhood involves turkeys. My grandmother was from a large Scottish family and had many brothers and sisters. One of her brothers was a farmer, and every Christmas he would send her a home-reared turkey for our family dinner. The dead bird would arrive by post with no wrapping at all, apart from an address label tied around its legs. It still had all its feathers intact, and I can remember the great excitement when the postman delivered it. Can you imagine that today?

There were no fridges in homes in the 1950s, so the coolest place to keep the turkey was in the bathroom (no central heating either), until it was ready for Gran to pluck it. So there it was, on a hook on the back of the bathroom door, just hanging around for Christmas. When I visited my Gran's during the festive season, I would never go to the toilet on my own, because when the door opened the poor turkey's beak would tap on the door and frighten the life out of me!

I obviously wasn't too upset, as I have always loved traditional Christmas dinner, especially my Gran's.

My first real pets, when I was about twelve years old, were two tortoises called Joey and Terry. My dad was very proud of his garden, and although it wasn't particularly large we used it regularly during the summer months, often having family and friends around for tea. This consisted of a variety of sandwiches and my mum's special homemade cakes with chocolate icing. We hadn't heard of barbecues in those days!

My dad built an enclosure in the garden for Joey and Terry and also made a beautiful painted tortoise house, with their names above the door, where they could shelter at night. However, we soon found out that tortoises can climb fences.

After several escapes, Dad reluctantly agreed to let them have the run of the garden. After that they roamed over the rockery and flower beds, not doing too much damage. Joey would run up the lawn (yes, really!) when he saw me approaching with his favourite treat, cucumber slices.

Feeding Joey the Tortoise

A tortoise's mission in life is to get beyond any fence or obstacle they encounter, as 'the grass is always greener on the other side' in tortoise world. Inevitably, they both eventually escaped from the garden, never to be seen again.

We also had two budgerigars while I was still living at home with Mum and Dad. The first one was called Bobby. He was a lovely bright blue colour and a real character. He only went into his cage to sleep at night and the rest of the time he had freedom to fly around the lounge. He had a play area on top of our large television set, with a swing, mirror and various squeaky toys. In the evening he would perch on Dad's glasses and watch the TV with him.

Our second budgie was called Ricky. He was emerald green and just as cheeky as Bobby. Ricky would fly into the kitchen to greet Dad, landing on his shoulder, when he heard him arriving home from work.

One day Ricky flew to meet Dad at the back door as usual, but was startled to see that he was carrying a large box. He flew straight over his head and out of the door. Dad ran into the garden, but could only watch helplessly as Ricky disappeared over the roof tops. We were devastated.

Dad took Ricky's cage and put it on the lawn in the back garden, just in case he came back. He then put lots of bread out to encourage the wild birds into the garden, hoping that Ricky would follow. Several hours passed and Ricky did not appear. It was now getting dark and we were almost giving up hope, but then, suddenly, there was Ricky sitting on a branch of the cherry tree in the middle of our garden. Dad went out, put his hand up and Ricky hopped onto his finger and quite happily went into his cage. A homing budgerigar!!

Toby and tragedy

When I was fourteen, after several years of pleading, I finally persuaded my mum and dad to get a dog. Mum was working part time now, just in the mornings, and I promised to take the dog out for a walk every day so they had no reason to say 'no' any more.

So Dad and I went along to Battersea Dogs' Home - I will never forget that awful day. We were told that there were no puppies available at that time, but we decided to have a look around anyway. We were shocked to see so many distressed dogs, gazing sadly through the bars of their cages as we walked by. I can remember thinking that the place smelt of damp and despair. It broke my heart and I came away crying - I wanted to take all of them home. I just couldn't get those sorrowful, soft faces out of my mind and had nightmares for weeks afterwards.

Dad reminded me that we had decided to adopt a puppy and, besides, we couldn't help all the dogs we had seen. He was right, of course, but I wonder now if that traumatic experience eventually led me down the road to animal rescue later in my life. I visited Battersea Dogs' Home again recently and, thankfully, there is now no comparison to the dismal place we saw in 1964.

Shortly afterwards, Dad heard some puppies were available at an anti-vivisection rescue centre in Watford. We went along and were introduced to a terrier cross, in a small dark kennel, and her three adorable puppies. The smallest one was black and brown and looked so sweet – it was love at first sight! So a few days later, we went back and adopted him. We called him Toby.

Toby was so small that he could sit in the palm of my hand. We took him home and he soon settled into his little bed – a cosy cardboard box with a soft blanket, in the corner of our warm kitchen. He seemed very quiet, but we thought he was probably just nervous in his new surroundings. However, it quickly became obvious that he was a very poorly puppy. He stopped eating after a few days and became very thin and weak. We took him to see the vet, who sadly shook his head and said Toby was a very sick little dog. He diagnosed distemper, a very serious, usually terminal, condition. I couldn't believe it and we were all devastated.

Me, 14 years old, with puppy Toby

I sat up with him all that night trying to get him to take some fluids. but he was fading fast. It became very difficult for him to breathe and the next morning, Mum and Dad took him back to the vet. I was too upset to go with them because I knew in my heart that my beautiful little puppy wouldn't be coming home.

We later found out that his mum and siblings had also died from that cruel disease. I cried for days and Mum and Dad said, 'no more dogs'.

The next dog that came into my life was Mickie, a black and white Border Collie who belonged to Joan and Brian, my sister and brother-in-law. They lived in a remote village in Essex called Colne Engaine, which at the time seemed a million miles away. We didn't have a telephone, so Mum used to write regularly to Joan every week and every Tuesday morning she would wait for the postman to deliver Joan's letter back with all their news.

Mickie

Mum, Dad and I used to drive up to Joan and Brian's every three weeks and stay for the weekend. Colne Engaine was very rural, to say the least, with just one bus service each week on a Friday to take shoppers to the nearest town, Halstead. I looked forward to our visits, as I never really enjoyed living in the hustle and bustle of the London suburbs and always longed to live in the country.

I loved to take Mickie out for walks around the surrounding country lanes, and I would pretend he was my own dog. I would chat to him as we walked along and tell him all my teenage problems. He would look at me with his big brown eyes, understanding every word. That imagination again!

Katy, the perfect dog

Our Katy

I had to wait until several years later, when I was married to Dave and living in Oxfordshire, before the next dog came into my life. Our children, Nicola and Daniel, were now six and four years old, so we decided that it was the right time to get a puppy. Dave wanted a large dog and I wanted one that didn't moult, so, after much thought and research, we decided that an Airedale would be ideal. The fact that they looked like big teddy bears was a bonus!

We saw an advert in the local paper for Airedale puppies in nearby Banbury, so we arranged to see them. They were all beautiful – five weeks old, little bundles of love and licks. One, a female, tottered over, gazed up and chose us.

Katy was a perfect dog. She never once messed indoors (even as a puppy), had a beautiful nature, loved the children and always did as she was told. She used to 'smile' a toothy grin at visitors. Once, when Nicola was poorly and the doctor came, she couldn't take her eyes off Katy, who was grinning at her across the lounge. Dr Terry kept repeating 'She's smiling at me, she really is smiling at me!' Meanwhile Nicola lay forgotten on the settee, patiently waiting for some attention.

Katy not enjoying being a mum!

When Katy was three years old, we decided to have puppies from her. As she was a 'big girl', we had to find a suitably smaller 'husband' for her so the offspring would be an ideal Airedale size. We took her along to a reputable breeder who had an ideal stud dog, but because he was smaller than Katy, he had to have a wooden box to stand on, with a mat on top to stop him slipping during mating!

The eventual result was nine beautiful puppies, but our Katy was not impressed. She found motherhood difficult, as she was such a laid-back girl, and found it hard to cope with nine hungry mouths to feed. We had to coax her into the box at feeding time, so we weaned the pups early for her sake. It meant more work for us, but Katy was happy.

We found suitable homes for the puppies all over the UK, the furthest going to Scotland. We decided to keep one, Gemma, assuming she would be as perfect as her mum. In fact she was the complete opposite – always in trouble, very naughty and chewed everything!

At this time we decided to sell up and buy a hotel in Torquay, but it wasn't really suitable for two dogs. Some friends who knew Gemma well asked if they could adopt her, so we agreed and life became much less stressful for us and Katy.

Katy adapted quickly to life in the hotel, in Wellswood, and was

Nicola + Daniel with Katy + Gemma

popular with the guests. She loved her walks through the woods down to Meadfoot beach, and lying contentedly in the garden. When we sold the hotel, Katy took it all in her stride and adjusted back to a peaceful home life.

Our new house was in a quiet close, and Katy would lie on the front lawn for hours, watching the world go by - or so I thought. I found out years later that she regularly trotted across the road to visit the neighbours, Ric and Marion, where she would be given treats.

Katy in the garden

Gemma - an independent little soul

When Katy was 12 years old, we decided to get a puppy. So Gemma II arrived – a six week old Lakeland Terrier. Katy was not amused, and they never really got on. We didn't appreciate at the time that it is not

Katy and Gemma

usually a good idea to introduce a puppy into the home with an elderly dog. Katy liked her peace and quiet, and she certainly didn't get that when Gemma was around. She tolerated Gemma, just.

Gemma was a real little character. We bought her from a breeder near Crediton, and the kennels were so filthy that we took her home the same day, although she was only six weeks old. She was so small

13

Puppy Gemma

that we couldn't leave her in our garden unattended for fear of a seagull snatching her for a snack!

Gemma was an independent little soul, or perhaps that should read 'naughty and disobedient'. She never liked children, and when she was just eight weeks old, she actually drew blood when she nipped a friend of Nicola's who tried to stroke her. When she was older, Gemma would lunge at any youngster we passed in the street, and she certainly didn't improve with age.

Despite going to training classes, Gemma never came when she was called. When she was let off the lead, she would take off across the fields and disappear over the horizon, eventually returning when she wanted to. However, we managed to tolerate her bad behaviour for over sixteen years.

Gemma in the snow

The puppy that chose me

George first came into my life when I was a volunteer for a local animal charity. The cats and dogs waiting to be rehomed by the charity were boarded at local kennels. I used to go to the kennels several times a week with another volunteer, Jean, to help clean out the cats and walk the dogs.

On this occasion, I took my camera with me to take photos of some puppies that were waiting to be rehomed. They had been born on a farm near Ipplepen, in the middle of a haystack, after their mum, who was a Beagle, got together with one of the working Collies on the farm. The farmer did not want the litter and threatened to shoot them when they emerged from the middle of the haystack. Luckily another volunteer, Jackie, heard about their situation and rescued all seven of the pups. She had saved their lives.

So there I was, in a kennel with seven eight-week-old puppies running around my feet, trying to get a photo of them for the charity's magazine. They were all different shapes and sizes, some with long legs and some with short. Some looked like their Beagle mum, while others resembled their Collie dad, and some were just a real mixture.

George, on the right, with his mum

Every time I looked down, there was one particular pup with long floppy ears sitting at my feet, gazing up at me with sad eyes. So I took the photos I needed and went home but couldn't stop thinking about my new little friend.

George, on the left, with some of his brothers and sisters

We had recently had our beautiful Airedale, Katy, put to sleep at the age of thirteen, but we still had Gemma, our naughty little Lakeland terrier. We hadn't planned to get another dog but when I told Dave about the puppy, he just said 'well, go back and get him then!' So I did, and George came home the same day.

George was quite a naughty puppy - he chewed my precious teddy's nose off, ate my Queen tape, then vomited the whole reel back up again and did the usual naughty things that puppies do. I remember one Boxing Day. The cold turkey was carved and plated up ready for our family dinner, and while I was out of the kitchen, George helped himself to all the white meat on the plate but left the dark meat. He obviously had good taste!

George was always special. He used to come with me when I did my voluntary work for the charity, and looked just like the charity's logo. He grew into a well behaved, quiet boy and everybody loved him. More about him later.

CHAPTER TWO

THE DAY THAT CHANGED MY LIFE

I first became involved with the charity Animals in Distress (Torbay & Westcountry) in 1990. My children, Nicola and Daniel, were teenagers and my husband Dave was working long hours in the fishing industry in Brixham. I loved my job working part-time in the Maternity Unit at Torbay Hospital, caring for the new mums and their babies, but felt that I needed something more to keep me occupied.

Dave and I had always been involved in working as volunteers for animal charities since we were married, including the World Wildlife Fund and Guide Dogs for the Blind. So I started looking around for an animal charity I could offer some help to. The main reason I approached Animals in Distress was that it was a local charity and the cat and dog logo was very appealing!

The charity was run from a garage in Ellacombe where second-hand goods were sorted and sold for both Animals in Distress and Guide Dogs for the Blind. I walked into the building, where Mrs Muriel Sellick, who owned the garage and was one of the charity's founders, and the treasurer, John Glaser, were sitting in a small, cluttered office, having a cup of tea. I uttered those fatal words 'Can I do anything to help?' That day changed my life.

I started helping in the garage by sorting the second-hand clothes, pricing them and putting them on to rails. The garage was large, but rather damp and cold. When it rained, the volunteers had to rush around putting up umbrellas to cover the rails of clothes, and place buckets in strategic positions to catch the drips.

There was a side room where Mrs Sellick and John would sort out the daily takings and discuss future events and ideas to raise the funds to pay the boarding kennel fees where the dogs and cats in care were temporarily homed. There was a small band of cheerful volunteers involved at that stage, and everyone had a passion for helping animals in need of care or rehoming. Committee meetings were held in members' homes, usually John's or Mrs Sellick's, and sometimes these meetings would become rather heated as they were an eclectic mix of personalities.

My favourite meeting place was definitely at the home which John shared with Bernard. Bernard was an excellent cook, and his home-made cakes and biscuits would be served with the coffee at the end of the proceedings. Scrumptious!

Dad and me helping Mrs. Sellick (behind stall centre)
at St. Marychurch Victorian evening

In the early years fundraising mainly involved coffee mornings in volunteers' homes or running a bric-a-brac stall at local events, the highlight of the year being the Babbacombe Carnival. Whatever the weather, the usual band of helpers would be there, sometimes trying to set up trestle tables in torrential rain and holding down the canopy to stop the wind whipping it away. We never gave up, as we knew that the amount raised, however small, would go towards helping the animals.

I often helped Mrs Sellick at various fundraising events, and I spent many a chilly evening with her at the Victorian Fair in the St Marychurch Precinct. She would carry boxes full of bric-a-brac and books which a grown man would have struggled to lift.

Mrs Sellick had great enthusiasm, so when she failed to turn up for a charity fair in Upton Church Hall one Saturday morning, I knew something was wrong. One of her neighbours and I went to her house and discovered her upstairs, lifeless. Typical of her, she had packed several boxes with books and knick knacks and they were stacked at the front door, ready to go. She was sadly missed.

Later, when the main cattery at the Animals in Distress Rescue Centre was built, we named it the Muriel Sellick Cattery, so her name is spoken every day – a fitting tribute to a remarkable lady.

I took over the role of voluntary secretary in 1992 and became more and more involved with the charity. I often accompanied John Glaser when he gave his talks to local groups, organisations and anyone else who wanted to listen!

I learned a lot from John. He truly loved the charity and was able to convey his enthusiasm and devotion to caring for unwanted animals. I hope I did the same in later years when I gave talks and slide shows. Many of my presentations were given to elderly folk in retirement and care homes, and they usually listened intently to my stories before spending the next hour or so telling me all about their own animals. They had treasured memories of their much-loved, long-gone pets.

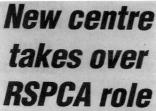

New centre takes over RSPCA role

HOMELESS animals were thrown a lifeline in South Devon today — days after a reluctant RSPCA confirmed the closure of its Kingswear base.

Torquay charity Animals In Distress is to take over a kennel complex in Ipplepen, giving hope to dozens of stray cats and dogs.

John Glaser, chairman of the 13-year-old charity, said today he was "delighted" that a year of negotiations had resulted in the purchase of Biltor Kennels, Ipplepen.

It was "a marvellous fluke" that the £2,000-a-month deal had been sewn up within hours of the announcement that the RSPCA's centre is definitely to close.

He said: "This will soften the blow — and it's a bad blow — to animals in South Devon. No cats or dogs will have to be destroyed.

"We are all devastated at the RSPCA's situation, but we will take up where they have had to leave off. We will pick up the slack."

New centre

The charity, which re-housed 700 dogs and 420 cats last year alone, takes over the building — to be known as the Animals In Distress Rescue Centre — on Monday February 21. It will house cats, kittens, dogs and puppies.

Mr Glaser said the centre would provide employment for "a small number of paid staff, but in the main we shall rely upon many volunteers to exercise the dogs and cuddle the cats."

The long-term plan is to shift the entire charity operation to the rescue centre.

Bosses hope income from their charity shop will cover the cost of the mortgage.

"We have always worked alongside the RSPCA. We are good friends with them, and we all work for one reason and that is for animals who can't help themselves."

Newspaper cutting

■ TO THE RESCUE ... John Glaser (right), chairman of Animals in Distress, with Dav Paradine (vice-chairman) holding pup George and Christine Paradine (secretary) wh will manage the new centre at Ipplepen.

Newspaper photo

I always tried to organise my presentations during the morning, as I soon discovered that following a good lunch and sitting in comfortable chairs in a warm, darkened room could send half my audience to sleep before I had finished talking!

I remember one incident when I became aware of a disturbance in the room as I was half way through my talk and slideshow to residents in a care home. A nurse went over to an elderly lady who had apparently had a 'wee accident' in her chair.

I carried on talking and clicking through the slides, trying to tactfully ignore the clean-up operation that was going on in front of me. I was offered a cup of tea and biscuits when I had finished, but politely declined the offer as the only available chair was rather damp!

In 1993 the charity received a substantial legacy, and finally the dream of running our own rescue centre became a reality. Volunteer

John Inman meeting visitors

Jean Rawson and I visited various banks and building societies to discuss the feasibility of the charity taking on a mortgage to purchase a suitable property. To our amazement, we calculated that the mortgage repayments would actually be less than the kennel fees the charity was paying at the time.

We contacted several local estate agents and visited many properties, none of which was suitable. After many months of searching, Biltor Greyhound Kennels in Ipplepen came to our attention. It was a rundown establishment and the kennels were in a bad state, but we could see the potential there so, in 1994, the Rescue Centre was created.

I left my job at Torbay Hospital and my family, George, Gemma and I moved into the bungalow on site, where I took on the role of Rescue Centre Manager.

The Rescue Centre was officially opened in September 1994 by the actor John Inman, who was appearing in the summer season show in Torquay. When asked if he would do us the honour of cutting the ribbon at the official opening, John naturally replied 'I'm free!' In fact he was absolutely free, as John did not charge the charity for his services that day, unlike several other celebrities who had been approached and asked for hundreds of pounds in appearance money.

After the opening ceremony, John spent a long time looking around the site and talking to the volunteers and visitors who had come along. He was happy to pose for photos, and he showed great interest in the animals and wanted to know all their stories. A true gentleman.

At first we had just two members of staff, twelve kennels and no catteries, but we did have many enthusiastic volunteers. The number of animals being taken in for rehoming or refuge gradually increased each year, so more staff were taken on to provide the care and attention required.

1. Building kennel block
2. Construction of the car park - mud, mud, glorious mud!
3. Extending reception area

Gradually, as funds became available, improvements were made and the Rescue Centre evolved into a cheerful, positive and rewarding place.

Christmas was always a special time at the Centre. Although closed to the public on Christmas Day and Boxing Day, the daily routine for the animals carried on regardless and there was usually the odd crisis or two to deal with as well!

All the staff came in on Christmas morning and were treated to hot mince pies and mulled wine after the animals had been cleaned and fed. Animal lovers would bring gifts for the cats, dogs, rabbits, rodents, birds

Animal Welfare Assistant Keith
giving Christmas treats

Santa (aka Fred Hammond) visits the
Centre with my Mum looking on

and stable animals leading up to Christmas and they were put under the Christmas tree in reception. In the early years, volunteer Fred Hammond would dress up as Father Christmas and hand out the treats to all the animals on Christmas morning.

Christmas 2005 was an especially busy time. The Centre covered for the South Hams dog wardens over the holiday period and at 5.30 pm on Christmas Eve a call came through from the police on our emergency mobile phone to say a stray Border Collie had been handed into Ivybridge police station and could I go and pick it up. When I arrived at the police station an hour later, a very frightened little brown terrier sat shivering in the corner of the police kennel – strange looking Border Collie, I thought. A police officer came out and informed me that a lady had lost her Border Terrier cross that afternoon and gave me a piece of paper with her name and telephone number. For some reason, the police did not think it was the same dog.

I put the little dog on the front seat of the Animals in Distress ambulance and he snuggled up to me, gave a big sigh and went to sleep! I phoned the lady from the car park opposite the police station and she described her lost dog – scruffy appearance, overshot jaw, and his name was Monty. My new little friend matched the description and opened

Handreared baby bunnies

"Little Willy Wonka"

Little Willy Wonka

one sleepy eye and wagged his tail when I called his name. I arranged to meet his very relieved owner and there was a happy reunion in the car park fifteen minutes later.

Unfortunately I had to take one of my own rabbits, Willy Wonka, to visit the vet on Christmas morning. I had hand-reared Willy and his three brothers and sisters from two weeks old, when their mother had died.

Willy was always the smallest and weakest and had a lopsided appearance, hence his name. He had now developed a chest infection, so he was put on antibiotics. I had to feed him liquid food through a syringe, as he was finding it difficult to eat with a blocked up nose. He was a lovely little bunny and overcame many medical problems in his short life.

On Christmas afternoon, as I was just dishing up the family's roast turkey with all the trimmings, the emergency phone rang again. It was Totnes police saying they had had a stray dog handed in and could it be picked up immediately. Dave generously volunteered to go to the police station, and his Christmas dinner was put back into the oven. When he was halfway there, another call was received to say the dog had been claimed by the owner. Hooray! Dave was back in 15 minutes and his Christmas dinner was still warm.

The same afternoon a lady brought a stray cat to the Centre. He looked very old and she was concerned that he might be ill. He was put into the warm cattery with some food (including a little roast turkey!) and snuggled up into his soft bed for the night. Around 5 pm on Boxing Day, a very worried lady telephoned the emergency mobile phone as someone had told her that they thought her cat might have been taken to the Centre. She cried with relief when I told her that he was perfectly safe.

Jessica, the Shetland pony

Apparently they had had visitors for Christmas who had brought their dogs with them and Truffles, who was 17 years old and never usually went out, decided he had had enough of the festivities and left home! The owner came straight up to the Centre and Truffles and his owner were reunited. There were hugs and tears all round, including me!

Waiting for a new home
for Christmas

At around six o'clock in the evening, the buzzer on the Centre's gates rang again and it was the wonderful people from the Ross Park holiday centre, just along the road. Every Christmas they would bring leftover turkey and ham from their restaurant, and they even took the trouble to dice it all into bite-size pieces. The dogs and cats loved it.

"Cool dogs"

Cool dogs

Once again the buzzer rang. This time it was a couple who had found a beautiful black Pointer type dog running around lost on Broadsands beach. I put the friendly but bewildered dog in a kennel and gave him some turkey and biscuits to cheer him up. An hour later, a frantic lady rang the emergency mobile and when I told her that her dog was safe with us, she burst into tears of joy. She and her family arrived at the Centre half an hour later and another happy reunion took place.

So, that was a typical Christmas at the Rescue Centre!

Snow at the Centre

Christmas humbug

CHAPTER THREE

RAISING OUR PROFILE

Fundraising always played an important part in the Rescue Centre's calendar, and my favourite event was the annual dog show. When we organised our first, about forty dogs entered and we were thrilled. However, over the years the event grew in popularity and numbers increased until nearly two hundred dogs were taking part and, one

An early Dog Show

year, visitors were actually turned away when the car park was full to overflowing! The majority of dogs entered had previously been rehomed from the Centre, so it was like a gathering of old friends with the proud owners eager to show off their perfect pooches.

It was always an emotional day, for all the right reasons, but none more so than when Milo the Rottweiler collected the 'Best Rescue Story' trophy. Milo had originally been adopted by Shirley Frampton, a regular volunteer and friend. Shirley suffered from chronic arthritis and was in constant pain, but never once did I hear her complain about her debilitating condition. However, she often did complain about everything else, as 'our Shirl' had very strong views on most subjects!

She was passionate when it came to animals, especially dogs, and much preferred them to human beings.

When her arthritis became so severe that she had to go into a nursing home, Shirley's two rescued Rottweilers, Milo and Lilo Lil, had to be rehomed, much to her distress. Sadly, Shirley passed away, but her family donated a trophy in her memory to be presented at our annual dog show.

Milo

Astonishingly, the first winner was Shirley's Milo, who had come along with his new owners. The show judge wasn't previously aware of the situation and Milo's owners didn't know about the trophy before they entered him in the class. A very special, emotional moment, and not a dry eye in the house!

Open Day

Me as a bunny pirate   Ready for Torbay Carnival

Torbay Carnival procession in Paignton

The Rescue Centre always had an Open Day each year, which was well attended by volunteers, members and visitors. Originally it consisted of a couple of trestle tables, sagging under the weight of bric-a-brac and books. Homemade sandwiches, delicious cakes and drinks were served from tables at the front of the garage, organised by Bernard, Sheila and a hard working team of volunteers.

This event evolved over the years to include many various stalls and marquees, live music, a licensed bar, barbecue and lots of other attractions to become one of the main fundraising events of the year.

During the 1990s, the staff and volunteers from the Rescue Centre took part in several local carnival processions to promote the charity. It was such good fun, and we all loved dressing up!

We always used a theme based on animals, of course, and many weeks of work went into creating our costumes and props.

Volunteers Sylvia and Ken played a major role in the preparations, and I still have the floppy, white rabbit ears Sylvia made for me. You never know when I might need them again!

The Sponsored Dog Walk from the Rescue Centre was always a successful fundraiser and one year, to try and make it more interesting for the participants, we decided to organise a quiz. The walkers would find the answers around the six-mile route. The receptionist, Geraldine, and I drove around the course the week before the event, but found it was a lot harder than we thought it would be to devise the questions.

We thought we were being very clever when we drove past Nutcracker Cottage and decided that the clue would be 'Where does the Sugar Plum Fairy live?' Brilliant!

On the day of the Sponsored Walk, the quiz sheets were handed out to the walkers and off they went. Unfortunately, unknown to us, the owner of Nutcracker Cottage had decided to paint his property on that very day and had removed the sign bearing its name. Not surprisingly, no one answered that question correctly!

Another popular fundraising event was the Dog Obstacle Race, in which the owner and their dog were sponsored to complete laps of an obstacle course. This proved to be a great fundraiser which everyone thoroughly enjoyed, and I will never forget the vision of a rather elderly lady valiantly tackling each obstacle but insisting on carrying her little poodle through the water jump so he didn't get his paws wet!

In the summer of 1997 the Obstacle Race proved to be especially memorable, being held on the wettest, windiest day of the year. At lunchtime we decided to abandon the stalls and other attractions, and a huge tarpaulin was erected over part of the car park to cover the barbecue and bar. If anyone turned up at least we could offer them food and drink!

Wet and windy BBQ

The obstacle course had been set up the previous day in the pouring rain, and the water jump was twice as deep as it should have been. Four o'clock arrived and the rain got heavier and the wind gusted to gale force, but the tarpaulin held up – just!

Then, to our amazement, people started arriving. There were eventually at least fifty cars in the car park and more than twenty dogs and their owners completed the course in the most appalling weather imaginable. Everyone was soaked to the skin, but we had a wonderful time and the burgers and hotdogs tasted better than ever. We even managed to make a small profit when all the sponsorship money was counted.

Publicity played an important role in the charity's success, and much to our delight, in May 1999, Animals in Distress was chosen from hundreds of animal charities across the country to be featured live in Thames Television's 'Pet Adoptathon' weekend. An advance party of producers and technicians had visited the Centre a few weeks previously and were very impressed with our set up. They wanted as many animal stories as possible, and the live broadcast was going on air from 7 pm, so a barbecue was arranged to ensure that a good crowd was present.

The television crew arrived at 8.30 am with outside broadcast vans, satellite dishes and enough equipment to fill our animal treatment

TV Broadcast John Glaser being interviewed

room, which was set up as their operations room. Hundreds of cables were laid, and by lunchtime, all was ready for rehearsals to begin. Westcountry TV presenter Alison Johns arrived and put everyone at their ease, as we were all getting rather nervous. Everything had to be timed to the last second, so each item had to be rehearsed over and over again. The afternoon passed very quickly and the crowds started to arrive.

Carol Vorderman and Matthew Kelly were presenting the programme from the studio in London, but as the broadcast was going out live, some items overran. As we were on last, our air time was gradually reduced. None of the dogs we had ready to feature were included in the programme and we were only on air for two minutes! However, it was seen by millions of viewers all over the country and the response was excellent, resulting in many of our animals being rehomed.

One morning I was very excited to receive a call from the BBC asking if I would like to take part in an outside broadcast from the Rescue Centre with local TV celebrity and weatherman Craig Rich. I readily agreed, as it would be excellent publicity for the charity and I would be a TV star for a day! A date was set for the transmission and I booked an appointment at the hairdressers for the day before – must be looking my best for my television appearance.

I didn't usually wear a lot of make up, but I did spend a few extra minutes in front of the mirror on the morning of our big day applying extra eye shadow, lipstick and even a little blusher. Craig had asked if he could walk around the Centre with me and we would chat about the work of the charity and some of the animals that were for rehoming – it sounded ideal.

I waited for the outside broadcast trucks to arrive with a mixture of excitement and nerves. A car drove in and Craig got out holding a tape recorder. I looked down the drive for the fleet of BBC vehicles, but, no, that was it. Craig explained that he was recording an item for his BBC Radio Devon programme. Oh dear, how embarrassing!

Fame at last

I hope I managed to disguise my disappointment as we walked around and talked about the Rescue Centre. Not quite the performance I had anticipated, but it was still good publicity.

Over the years, many church services involving animals have been organised to support the charity. The first took place at Furrough Cross in St Marychurch, Torquay. They were always, and still are, well attended by many owners and their pets.

After the Rescue Centre was opened in Ipplepen, the village church held several services for the animals in a field adjoining the church hall. We always seemed to be lucky with the weather, and I can remember one year when the congregation and their various pets shared the field with some very curious young bullocks. They didn't come too close, thank goodness, but watched the proceedings with great interest.

Chicken in Church

Henrietta Rabbit in Church

Annual animal thanksgiving services have been held in Totnes, and an amazing variety of pets have been brought along over the years including dogs, cats, rabbits, guinea pigs, rats, hamsters, tortoises and one very brave chicken! They all behaved reasonably well and the highlight was always when some of the dogs joined in with the singing of the hymns.

Diary of an animal rescue centre

During my time at the Rescue Centre, Animals in Distress was the only animal charity in the area to provide a 24 hour emergency telephone service that was always answered by a real person (usually me!) and not an answer machine. All the appeals for help were logged and these are just a sample of some of the genuine phone calls received:

1st January, 7.30am. Man looking after daughter's rabbit over Christmas. Run out of food – shops shut. We will supply some to him, on his way now.

20th February, 8.15pm. Feral cat trapped in lady's roof. Advice given re leaving food out, etc. If cat trap needed, will ring Centre in morning. (No further contact, so presume cat came down!)

20th March, 6.00am. Lady had dog from us two days ago – run away. Very upset.

20th March, 6.10am. Found dog down the bottom of garden. Panic over!

29th March, 8.00pm. Lady's cat being sprayed with water from next door neighbour. Advised to discuss with neighbour.

14th April, 7.00 pm. Ivybridge Police Station. Man deceased. His young GSD now in police custody. They will keep it kennelled overnight and we will pick up am for rehoming.

18th April, 9.30pm. Lady's dog rejected one of her puppies. Could we take in and hand rear? Will bring puppy over now.

Granddaughter, Jazzy, meets puppy

29th April, 10.25pm.   Man ringing from Wales! His son's parrot has escaped and is now sitting on their roof. Advice given.

2nd May, 7.15 pm. Children playing in park found box of four baby rabbits. Will go and collect now.

Baby bunnies found in a box

7th May, 9.15pm. Lady has had house fire this evening. Can we take her cat and 3 week old kittens? She will bring them to Centre now. Waited until 11.30 pm, then lady rang again. She has made alternative arrangements.

12th June, 8.15pm. Owner of local pub can hear kittens crying in wall cavity of disused building. We will go now to investigate.

12th June, 10.30pm. Pub landlord rang to thank us for picking up the kittens and their mum. Wants an AiD collection box for his bar, please!

20th June, 2.30 AM! Lady rang. Neighbour's dog barking, keeping her awake. Can I do something? Advised to discuss it with her neighbour in the morning!

Kittens found in a shoe box

10th July, 6.45pm. Man found cardboard box with two tiny kittens in St. Marychurch. Will bring to Centre now.

8th August, 8.45pm.   Lost puppy. Black Labrador, 4 months old.   Fell into River Dart at Newbridge this afternoon. Details taken. Lady very upset.

9th August, 7.30 am. Puppy found! They went back to search riverbank with torches last night and found him hiding in undergrowth – hooray!

12th September, 8.30pm. Lady rang. Her dog has gone into labour in her father's bedroom. Gave advice and talked lady through the birth of first pup.

12th September, 9.15pm. Lady rang again. 3 puppies now arrived. Everything seems fine. Will ring Centre am.

13th September, 8.00am. Final count – 7! Mum and pups doing well. Lady's father not very happy though! Thanks for help and advice.

These were just some of the calls I received out of hours. Many were not real emergencies, just anxious people needing advice or reassurance concerning an animal, and that is what the charity provided at that time.

CHAPTER FOUR

DOGS ARE FOR LIFE

To describe a typical day at the Rescue Centre would be impossible, because they were all so different. When the phone rang or a car pulled into the car park, we never knew what situation we would be dealing with. A box of tissues was always on hand in the reception area, as emotions can run high when animals are involved, happy as well as sad.

My life at the Rescue Centre was always busy, sometimes emotionally draining, but never, never, boring!

While we were living there, Dave and I shared our home with animals of all varieties, shapes and sizes, particularly, of course, dogs. In this chapter I would like to introduce you to some of the canine friends we made.

George

George, my constant shadow

My own Collie cross, George, was always by my side, my shadow. He never went on a lead and was very gentle and patient with the many animals that came to the Centre for rehoming, including kittens, chickens, rabbits and the baby goats, Harry and Alfie.

George and kitten

George took life very seriously and watched all the comings and goings with a thoughtful expression. He did have a sense of humour though, and would smile (yes, really!) showing all his teeth when something made him happy.

George loved everyone – apart from vets! The charity's vet visited the Centre twice a week and George was the early warning system of his arrival. He would start barking, nonstop, before the vet was even on site - somehow George always knew he was driving up the lane. I would have to stop whatever I was doing and take him indoors, as his barking got louder and louder as the vet drove up the drive and into the car park.

After the vet had left, George returned and, with nose fixed to the floor, would trace every step where his antagonist had walked, just to check he was off the premises. Only then would George relax and return to his usual laid-back self.

The following is an article I wrote for the charity's magazine, entitled A Day in the Life of George, in his own words:

'*My day usually starts far too early, at around 4.30 am when Dad gets up to go to work. I usually pretend to be asleep in my basket, but he insists that I go outside into the garden, whether I need to or not. If it is cold or wet, then it is a quick dash to the lawn and back, to keep him happy, then I can settle down for another couple of hours' sleep before breakfast.*

Life is generally pretty busy for me at the Rescue Centre. It is my responsibility to check that the inmates' food is prepared properly by the staff and, of course, I have to clear up any meat or biscuit dropped on the floor and any food bowls left around have to be checked for leftovers. Bearing in mind that I have only just devoured my breakfast indoors, it is sometimes quite a chore to eat every scrap but somebody has to do it.

Once my kitchen duties are done, I trot down to Reception and listen in to find out who is arriving today and, best of all who is leaving for their new home. I don't get to know many of the cats, as they usually arrive in baskets and I am not allowed into the catteries in case they are frightened of me - can you imagine that!

Sometimes, we have orphan kittens indoors for Mum to bottle feed and I let them snuggle up to me as they smell lovely and milky! I like to help by licking them clean at both ends!

A scruffy dog has just arrived with his man and is walking happily, wagging his tail, from the car park into Reception. When Keith is called over from the kennel block, I know this means the dog won't be leaving with his owner. He trots off down to the kennels, unaware that his life is about to change forever. The man walks back to his car, wiping his eyes. I wish I could tell him that most of the dogs that come in don't usually stay too long and they have lots of lovely walks and the people here will make a fuss of them but, it is too late, he has driven out of the gates.

Now it is time for a nap. My favourite spot is on the landing, half way up the stairs to Mum's office. There, I can hear what is going on in Reception and also make sure that Mum doesn't leave her office without me. I can also check on visitors as they step over me. I may sound as if I am snoring but I'm ready for action, really!

I have seen two dogs go off to their new homes today. Molly the Jack Russell was looking out of the window of the car, grinning from ear to ear, sitting on the lap of her new friend. She winked at me as they drove past - I think she has them trained already! The big Collie cross, Max, looked anxious as he walked to the car park with his new man. He has already lost two owners, through no fault of his own, so he is going to try really hard to hold on to this one. Hope I don't see him again unless it is just for a visit.

Great, it's dinner time. I supervise in the kitchen, as usual, leaving the floor spotless. How would they manage without me?

George was my best mate and everybody's friend (apart from the vet!)

George playing with pups

Bruno's mum Rosie and her pups

Bruno

'Not the brightest of dogs'

Bruno's mum, Rosie, was pregnant when she was brought in for rehoming and gave birth to her nine pups at the Rescue Centre.

Rosie was a Rottweiler and her partner was a Springer Spaniel, but the puppies all looked just like Labradors. Bruno was the largest in the litter and I was there to provide a helpful tug to pull him out as he was such a big puppy!

He was a sweet, affectionate boy but not the brightest of dogs and his favourite pastime was chasing his ball in the field, nonstop. He was overweight due to a thyroid problem (as we found out too late) and he sadly died when he was only 8 years old.

CHAPTER FOUR

Slobberchops

Storm (named because he was born during a thunderstorm on Dartmoor) came to the Rescue Centre for rehoming as his owners had divorced.

A large, long-haired St. Bernard, Storm was a one-man dog. He never really liked me, apart from meal times! The first time I saw him, in the Centre's car park, he growled a warning but when Dave approached him, Storm slowly wagged his tail and their friendship began.

Storm was good with children, but despite his size he was never 'top dog' in the household. He used to be bullied by our German Shepherd, Sheba, and she would ambush him in the garden by jumping out from behind a bush and send him running off with his tail between his legs. Storm was so big that he ate and drank from horse buckets.

Storm and grandson, Will

Unfortunately, St Bernards are real slobberchops, and after he had died, we were still finding trails of Storm's slobber for many months in the most

Storm relaxing with his friend, Tia

unlikely of places, for example, on the ceiling and on a framed photo on the wall of my Dad, who wasn't his greatest fan!

Susie and Duke

Gemma and Susie

Susie was a sweet, cross-eyed Airedale. Airedales are my favorite breed of dog, and when Susie came to the Centre for rehoming she came straight indoors.

Airedales are usually known for their intelligence but not our Susie, bless her! Her passion in life was chasing her special ball, which she would do all day if allowed to.

Duke

Duke was the first of our two German Shepherd dogs. He was quite elderly when he came to the Rescue Centre for rehoming and had severe hip dysplasia. This is a painful condition that is quite common with GSDs and affects their ability to walk on their back legs. He was a lovely, docile old boy and got on well with our other dogs.

Lurch and Sandy

Lurch

Lurch was an ancient Lurcher who had had cancer and was not in good health when he was brought in for rehoming. We brought him indoors to live with us as he was so unhappy being in a kennel and didn't have long to live. He lived on for a few more months until another tumour appeared in his neck so he was sadly put to sleep.

Sandy was a 14-year-old Collie cross who came to the Centre for rehoming when her owners moved but couldn't take her with them. She had mammary tumours, which were successfully removed, and eventually moved into our home as nobody else wanted her. She was the only dog we had that was ever allowed to sleep on the end of our bed! After three months, her health deteriorated quickly as the tumours returned and she became irritable and stressed. She left us very peacefully but left a big hole in my heart.

A gentle giant

Fenn was the first of our two Great Danes. She came indoors, as Great Danes find it very hard to cope with life in kennels, so she was very distressed. She was a gentle giant and got on well with George.

Fenn lived with us for a few months, and then a lady came along who wanted to adopt her. Following a successful home check, Fenn went off to her new home. We missed her a lot and felt quite sad when she left, but thought it would be best for her to be in a home where she was the only dog, with lots of fuss.

Great Dane, Fenn

Several years later, much to our dismay, Fenn was returned to the Rescue Centre as her owner said she could no longer afford to keep her. We were shocked when we saw the condition she was in. Fenn was very thin and could barely walk, so she was taken straight to the vet for a check up. The vet told us that she was a very sick dog and in a lot of pain, so the decision was made to have her put to sleep.

To this day, I still feel guilty and wish that we had kept Fenn ourselves but, in life, you have to make decisions that you think are right at the time although they don't always turn out to be.

Handsome Boy Basil

Banned from the local

Basil, the Great Dane, lodged with us for many years. His owner, Claire, was a volunteer and would always bring Basil with her when she came to help at the Rescue Centre, so we knew him well. When Claire decided to work abroad, she asked if we could look after Basil until she returned.

So Basil moved in and Claire would come home every few months and visit him. I used to write letters to her as if they were from Basil, with all the charity's news. He was a real character and was a regular

Basil and Bruno tug o war

at the Wellington Inn in the village with Dave, until he was banned because of his wind problem - Basil, that is, not Dave!

As he grew older Basil's health wasn't too good, but he was always so pleased to see his mum when she came home for a couple of weeks' holiday. On her last visit, Claire took Basil out for a walk along the footpath by the River Dart on Dartmoor, one of their favourite places, but he had a heart attack and died in her arms. She arrived back at the Centre, very distressed, with his body in the back of her car. Basil was buried at the bottom of the Centre's field and I always think that he waited for Claire to come home so she would be with him at his end.

CHAPTER FOUR

A dog from a broken home

Sky

Sky came in for rehoming when her owners divorced. She was a two-year-old French Mastiff crossed with a Rottweiler. She was so miserable and frightened in the kennels that nobody could get near her – she just sat and growled and showed her teeth to everyone!

Dave saw her and it was love at first sight! He sat on the floor in the kennel, just ignoring her, for hours and gradually she came across and ended up sitting next to him. She was successfully introduced to our other dogs so we took her indoors and she was like a different dog. Sky is still with us, at the time of writing.

Tiko, the wonder dog

Tiko, a Long Haired Dachshund, arrived at the Rescue Centre when his owner died. He was 14 years old but a healthy, bright little boy. It would have been difficult to rehome him because of his age, so he came indoors. He followed me around the Centre during the day and was a happy little soul.

Tiko the Wonderdog

Why was he called Wonderdog, I hear you ask? Well, one day Tiko was following me around as usual when a car came up from the car park and didn't see him. The back wheel of the car went right over him and he screamed with fright and pain. One of our receptionists, Sue, drove me to the vets with Tiko on my lap wrapped in a blanket.

I didn't know if he would still be alive by the time we got there, so I stroked him gently, but he (and I!) cried all the way. He was rushed into the vet's surgery and they said they would x-ray him to see what damage had been done and if he could be saved.

Then a miracle happened! After I had laid him on the table, Tiko stood up, shook himself and looked at me as if to say, 'what's all the panic about?' His x-ray showed bruising in the shape of a tyre mark right across his stomach, where the wheel had gone over him, but that was all the injuries he had. He was sore for a few weeks and walked very slowly, so we built him a special step so he could get in and out of the back door more easily.

Tiko went on to make a full recovery and lived to the ripe old age of 16 years.

The dog that was left behind

Tia the Bassett Hound came to the Rescue Centre for rehoming with her best friend, a King Charles Spaniel called Honey. They had been rescued from breeding kennels in Wales, but when their new owner

Tia and Honey with volunteer, Miriam

died, they were brought in for rehoming. Tia was around eight years old, but Honey was older and had heart problems. They were rehomed twice but returned, first because Tia apparently snapped at a child and then because they weren't house-trained. They had spent most of their lives living in kennels, so this was to be expected, and the new owners were aware of this, but they still brought them back to the Centre.

As they waited patiently for new owners to adopt them, Honey's health started to deteriorate. She had tests at the vets which showed that she had a serious heart condition, which is quite common in King Charles Spaniels. Sadly, she suffered a heart attack. I will never forget Tia's expression on her face as she watched her best friend being taken away, wrapped in a blanket. Honey didn't come back.

Tia cosy with a friend

I couldn't leave Tia in the kennel on her own, so she came indoors and lived with us for the next five years. She wasn't house-trained, but she never showed any aggression and used to follow me around the Rescue Centre. She was very popular with visitors and soon made lots of friends.

Tia developed an eye condition called glaucoma, which is common in Basset Hounds, and lost the sight in one eye. The eye had became very swollen and painful, so the vet suggested that she should have it removed. She coped very well with the operation and with only having one eye. Unfortunately her other eye also developed glaucoma and she lost the sight in that one too. The vet removed the eye, and I didn't know whether it would have been kinder to have her put to sleep, but decided to give her a chance.

She coped well with being blind, got on with life without a fuss and lived until she was 13 years old. My little heroine.

Saved from squalor

Honey the Rough Collie was a very special dog, and I will never forget the day she arrived at the Centre. Her owner was an elderly lady who had become a recluse. She lived in a bungalow, and Honey had never been outside in the three years of her life. When the lady was taken ill, social workers arrived at the bungalow and were shocked at what they found. The whole place was filthy, with mounds of rubbish and dog mess in every room. Honey had never even been into the garden, so there were layers and layers of soiled, wet newspaper. Honey's food was just thrown on to the floor – she had never eaten out of a bowl.

Honey

When the social worker arrived at the Rescue Centre, Honey had to be carried from the car to her kennel, as she had frozen with fear and could not move. For several days she had to be carried on to the grass to do her business. She had never worn a lead, so this too was a frightening experience for her.

We found a home for her quickly, but received a phone call the next day telling us that she was lying on the settee in the lounge crying, and would not move. Unfortunately, her new home had wooden flooring and Honey would not walk on it. We could hear her howling in the background when her new owner phoned to say she was bringing her back.

So Honey came indoors with us. We had carpets in the hall and lounge where there was a door into the garden, so that was not a problem. She wouldn't go into the kitchen on the lino covered floor – I think it was the noise of her own claws on the floor that frightened her.

At first I fed her by hand, but gradually she got used to eating out of a bowl, like our other dogs. I got her used to having a lead put loosely around her neck, and eventually she went out with us through the front door – only a few paces, though. Any loud noise or sudden movement and Honey would panic and rush back indoors. After several weeks, she could walk down to the field with our other dogs and would run around with them. It was a lovely sight to see her prancing around, carefree and happy.

Getting her into the car was the next challenge. Sky and Tia would jump in and then we lifted Honey in. She wasn't happy at first but soon got used to it. It was a great day when we took them all to Goodrington Sands and Honey walked along the beach on her lead, for the first time. What an achievement for her (and us!).

Honey then started having fits. They would last for up to 20 minutes, and she was very distressed. Sky would sit by her side until she stopped shaking. They became more frequent, so she was put on medication, which kept the fits under control. We noticed that she also seemed to have problems with her sight, so I took her to the vet eye specialist in Tavistock, and he confirmed that she had very

limited vision, which was why she moved her head from side to side a lot. It was a condition called 'collie eye' and she had been born with it. Unfortunately, there was no treatment available.

As time went on, we got her to walk round the village on her lead next to Sky. Any loud noises still made her jump, but considering all her problems, she coped well.

When we left the Rescue Centre and returned to live in Torquay, we knew Honey would not be able to cope with life in a busy town, so I

Sky and Honey

contacted Rough Collie Rescue and they found her an ideal home in Somerset. It was a very sad day for me when we took Honey to meet her new owner, but they took to each other immediately and both seemed very happy when we left. Her new owner phoned several times to tell us that Honey was fine.

Refugee from violence

Sheba was a dog who truly touched my heart. We received a phone call at the Rescue Centre from a lady in Torquay, who was very upset. She told us that she was going into a women's refuge with her children, as her husband was being violent towards them. They had a six-year-old German Shepherd, Sheba, who had been neglected and needed to be rehomed. She said she would call us back when her husband was out, so we could go and pick Sheba up, as he would never agree to it. A few days later, we received another call from the lady saying 'come round quickly, he has just gone out', so two of the kennel staff went to the house immediately to pick up Sheba.

When they arrived back at the Centre, it was a terrible shock. Sheba was emaciated, had little fur left and her body was covered in sores. She had eye and ear infections and her claws were so long she could barely walk. It broke my heart to see a beautiful dog in this condition. She looked so sad and had a haunted look in her eyes. There were many tears shed that day.

I contacted the local RSPCA Inspector, as this was a case of cruelty. Even the Inspector was shocked when he saw Sheba, and

The day Sheba arrived

took photos of her as evidence. The owner was later prosecuted, fined and banned from keeping a dog for ten years. A small price to pay for inflicting such suffering.

Sheba

My beautiful Sheba

George and Sheba

So Sheba's long road to recovery began. She had such a sweet nature, despite the suffering she had endured, that I couldn't help but fall in love with her. Her mange was treated by regular baths and she walked much more easily once her claws had been clipped. She was half the weight she should have been, but soon started to gain strength once she started eating regularly - little and often.

George and I spent many hours in the field with her that summer, especially in the evenings. She was very shy, so I used to sit on the grass and gradually, as I gained her trust, she came over and would eventually sit quietly next to me.

Sheba had never been socialised and didn't know how to play. I started to throw a ball for her and she would just watch it and then look at me as if to say, 'now what?' I will never forget the day when she chased the ball and brought it back to me for the first time – I cried tears of happiness!

Once the mange had cleared, we successfully introduced Sheba to our other dogs at that time, Bruno, Storm and Gemma, and she moved indoors with us. She had a wonderful nature and loved lots of cuddles. Unfortunately, she had hip dysplasia which meant her back legs were weak and this gradually became worse as she got older. Sheba was with us for about five years - another very special dog.

The puppy that couldn't see

Puppy Paxton

Ah, little Paxton. I first saw Paxton's photo on the Many Tears charity's website. He was a six week old Basset Hound and was blind. He had been born on a puppy farm in Wales, but because of his condition was not considered suitable for rehoming, so was given to the animal charity.

I already had Tia, another Basset Hound, who was blind and now also deaf. Tia was a happy little girl who accepted her disabilities without fuss and so I felt I could offer Paxton a home, with the support he would need for the rest of his life. And he was so irresistibly cute!

So Dave and I travelled up to Wales to collect this gorgeous little puppy. His foster mum met us in a car park, and as we drove in I could see this tiny bundle in her arms and it was love at first sight – in fact, I burst into tears! We followed her car back to the Many Tears kennels, where the adoption paperwork was completed. It was a three-hour drive home and Paxton curled up in my arms and slept most of the way. We stopped once at a service station on the M5 where he did a pee and a poo but didn't mess once in the car.

When we arrived home we introduced him to Tia, who didn't seem too impressed, but Paxton loved her and tried to get into her box with her. He eventually managed to squeeze in when she was asleep! Paxton soon found his way round and slept right through the night with the other dogs, Sky and Honey, with no fuss. Housetraining wasn't easy, but he soon learned where the back door was into the garden and that this was where he was supposed to perform his duties. He came to my office with me every day and settled down quietly in his box while I was working.

My granddaughter Emily and Paxton

Paxton snuggling with Tia

Paxton has grown into a confident, well balanced little dog. He gets a lot of attention when we go out and is good at meeting new people and dogs. He can run off the lead in a secure area, but never wanders off far. He knows that when I clap my hands it means 'come', which he does - most of the time. If he is running towards a hazard such as a fence or tree, I shout 'stop' and he stops. When we are walking along a road, I say 'step up' or 'step down' at the kerbs, so he knows what is ahead and lifts his paw up to overcome the obstacle.

When Paxton had his first trip to the beach, he typically took it all in his stride. When his paws first felt the cold sea-water he stopped in his tracks, had a long sniff at the lapping waves, then carried on

Paxton

regardless. Goodness knows what he must have been thinking, but because Sky was there to guide him, he accepted the new experience without fear, as he does most things. It is sometimes hard to believe that he has no sight all. Just like Tia, Paxton is a shining example of coping with a disability.

Many animals shared our family home during my seventeen years at the Rescue Centre. All had their own stories to tell and all have a special place in my heart.

It would be impossible to mention all the dedicated volunteers and staff who have been involved with Animals in Distress over the years, but most importantly, it has always been about the animals. Since the Rescue Centre opened in 1994, thousands of animals, including hundreds of dogs, have been cared for and rehomed.

A doggy love story

Two German Shepherds, Chloe and Jasper, arrived together at the Rescue Centre for rehoming. They were inseparable, so they would have to be rehomed together. However, it soon became apparent that Chloe was expecting puppies. As her time came nearer, she was moved to a special puppy kennel on her own so she would have comfort and safety for her new offspring. That night I was awoken by the sound of howling and, on investigation, I discovered that it was Jasper and Chloe communicating with each other. A relieved Jasper was moved to the kennel opposite Chloe, where they could see each other, and peace was restored.

Chloe and her pups

Chloe gave birth to nine beautiful puppies, seven boys and two girls. Jasper was the proudest dad in the world and watched all the comings and goings in Chloe's kennel with keen interest. The pups grew quickly, and when they were weaned and old enough to be away from their mum, Chloe and Jasper were reunited, much to their delight. Jasper had been neutered by now and Chloe was spayed a few weeks later.

The lovebirds were once again inseparable and the search to find them a home together began. Their puppies had all been found new homes, but Chloe and Jasper waited patiently in their kennel for someone to come along and offer them a forever home together – and, of course, they did!

We all love happy endings, and Chloe and Jasper found theirs. When they visited the Rescue Centre with their new owners a few months later, it was obvious that they were all very happy together, and the love story continued.

Puppies in their foster home

Three six-week-old Cocker Spaniel puppies were rescued from a puppy farm, as they had problems with their eyesight and were going to be put to sleep by the breeder. When they arrived at the Rescue Centre, the pups were immediately placed with

Puppies in their foster home

foster carers, Sylvia and Ken, who regularly looked after puppies for the charity in their home until they were ready for rehoming.

Following an examination by the vet, we realised that all three had very serious eye problems, so they were referred to a specialist vet in Tavistock. The pups, now christened Milly, Molly and Milo, were wrapped up warmly in a blanket ready for their long trip in the charity's ambulance. Bearing in mind their age, they were the perfect passengers, and all three were clean during the journey across Dartmoor to the vet's surgery.

They were diagnosed with multiple eye defects – Milly having the least sight, Molly limited vision and Milo limited vision in one eye. That was the bad news – the good news was that they did not need surgery and their sight would hopefully improve as they grew older. All three pups were rehomed to families who understood their special needs, and their owners all reported that they were doing well.

Bouncing Harvey

Harvey, a 10-year-old Dalmatian cross, came to the Rescue Centre

for rehoming, as his owners could not cope with him. He was very boisterous, and being a big boy, he took a lot of handling. Harvey took great delight in barking at visitors and jumping around his kennel with excitement like a lunatic. Unfortunately this didn't exactly endear him to potential owners, and people would pass him quickly by. Notices were put on his kennel informing the public that this wasn't his normal behaviour and that out of the kennel environment his manners would improve.

Harvey

Two years passed and Harvey was still waiting for his special home. Then, finally, a couple visited and saw through his bad behaviour. They visited the Centre several times, taking him out for walks and getting to know him. When they eventually took him home he soon settled in and even had his own couch to sleep on.

CHAPTER FOUR

The sheep chaser

Hazel was a young brindle lurcher who was brought in for rehoming as she lived on Dartmoor but had an irresistible urge to chase sheep. She had been at the Centre for eight days when a new home was found for her with a lovely, caring family who lived in Torquay. Perfect - no livestock in sight!

Two days after she had gone to her new home, she suddenly bolted out of the open front door and was gone. Her new owners were desperate and, after a search for her, phoned the police, dog warden and local kennels. Three hours later, Hazel calmly walked up the drive to the Rescue Centre, wagging her tail. How she found her way, approximately eight miles, back to the Centre, and indeed why, we will never know. Her relieved owners came to collect her immediately. Hazel was very pleased to see them and they had no more problems with her.

A reason to get up in the morning

Dolly and Daisy with Animal Welfare Assistant, Carol

Dolly and Daisy, two 10-year-old grey Miniature Poodles, were brought to the Rescue Centre for rehoming after their owner had died. Dolly was the leader and the stronger of the two, as Daisy had cataracts and was completely blind. Dolly was her 'guide dog' and Daisy would follow her everywhere.

There was a specialist eye vet in Tavistock, so Daisy was booked in for an appointment to see if anything could be done to give her back her sight. Dolly went with her, of course, as they were inseparable. The vet said that both the cataracts could be removed and some of her sight could be restored. Wonderful news! Daisy was booked in for her operation the next week. Everything went well and Dolly came with me, in the animal ambulance, to collect Daisy from the veterinary surgery following the operation.

On the way back, I was driving across Dartmoor with the two of them curled up on the seat beside me. Suddenly Daisy slowly stood up and looked out of the window! What a moment for her – the first time she had been able to see for many years.

She gradually gained more sight as the weeks passed and her life had been changed for the better forever. They were adopted by a lovely lady called Sandra who sent us regular progress reports. The following is taken from one of Sandra's letters:

'They have been a sheer delight from the start. I do not think a day goes by without them making us laugh, they get up to such antics. They are the most loving little girls one could ever come across.

'They have helped a 10-year-old boy overcome his fear of dogs. Thomas came to visit, but when he saw the dogs, he literally froze with fear. I promised him that they would not hurt him and he reluctantly came through the door but tried not to come into contact with them. He sat down in a chair and suddenly Dolly was on his lap! She was so good with him. Thomas's mother couldn't believe her eyes. By the time they left to go home, Thomas had both Daisy and Dolly each side of him as if he had known them all of his 10 years. I was meant to have these lovely little dogs, if only for that. '

She finished her letter, 'Dolly and Daisy have given me a reason to get up in the morning. Thank you, thank you for allowing us to have them!''

When Rudi met Sally

One Christmas, two young Lurchers were brought into the Rescue Centre by the police, having been abandoned. Following a domestic dispute involving their owners, the dogs had been locked for several days in a flat with no food or water. Thankfully, worried neighbours alerted the police to their plight. It was not certain how long they had been there, but the only way they had kept themselves alive was by drinking the water from the toilet, as luckily the loo seat had been left open.

The pair were in poor condition when they arrived at the Centre and were put in the warmest, most comfortable kennel together. Being Christmas time, the staff named them Rudolf and Blitzen.

Rudi

They were so malnourished that Rudi had to be put on a drip at the vets and in fact 'died' several times during his treatment, but eventually he managed to pull through. Following lots of TLC and regular meals, the pair eventually made a good recovery and their individual personalities began to shine through. Rudi was the clown, while Blitzen was the quieter of the two. They found new homes individually, but Rudi, because of his antics and bad behaviour, was returned on several occasions.

Sally, a regular supporter of the charity, had adopted several dogs from the Centre in the past, usually the most needy ones. When she saw Rudi gazing at her from his kennel with a silly expression on his face, he made her laugh and it was love at first sight. So Rudi went off to his new home, and we knew he wouldn't be back this time. Luckily for us, Sally liked a challenge!

However, that was not the end of Rudi's traumas. About a year later, he was playing in the field adjoining Sally's home with one of her other dogs, Max. Suddenly, she heard a piercing scream and Sally ran down the field to the long grass where Max was standing. She found Rudi lying on the ground motionless with a gaping hole in his chest. It was so deep that Sally could see his exposed heart beating.

A neighbour helped to lift Rudi into Sally's car and he was rushed to the vet. Despite being the most severely injured dog ever seen at the surgery, Rudi managed to survive. Over the years that followed, the accident-prone Rudi has had so many stitches inserted by the vet to repair various injuries that he now resembles a patchwork quilt!

Helen Chamberlain

Helen and Spanner

Sky Sports presenter Helen Chamberlain was a loyal supporter of Animals in Distress for many years, and on several occasions she stood on the streets of Torquay rattling a collection tin with Mrs Sellick, John and Bernard. Helen had never owned a dog before, being 'strictly a cat person', but during a visit to the Rescue Centre she discovered a little brown terrier called Petal tucked away in our isolation kennel, with not much fur left and covered in mange. She couldn't get that little mutt out of her mind, and on her next visit, a few weeks later, Helen adopted Petal, renaming her Spanner.

For seven years the pair of them were inseparable. Spanner turned out to be an intelligent, perfect companion and they often visited the Centre to take some of the other less fortunate dogs, still waiting for new homes, out for walks.

Elvis and Spud.

Elvis and Spud

During one of their visits a van arrived from CARA, an animal charity based in Wales. We often helped them out when we had room in our kennels by taking some of the dogs rescued from council pounds who were desperate for rehoming. For some, we were their last hope.

On this day, when the back door of the van was opened, we were greeted by the usual distressing sight of numerous wire cages piled high, containing various breeds of beautiful dogs, all looking at us with pleading eyes. An ancient Jack Russell terrier rolled out of the van on to the ground, picked himself up and stood there with his little legs shaking, looking up at Helen. That was it. Elvis went home the same day with Helen and Spanner!

When Spanner was just seven years old, she developed cancer and within four months, a devastated Helen telephoned to tell me that she had lost her fight for life. So when a cheeky eight-month-old Jack Russell came

Spud - how could you resist him?

into the Centre for rehoming, I sent Helen an email with his photo attached. I knew she wouldn't be able to resist his doggy charms, though he certainly wasn't the best behaved, or the quietest, dog on the planet.

Helen worked hard with Spud over the following months and his behaviour did improve - a little. However Helen would never part with him, despite him being a pest and constantly terrorizing her hens and turkeys.

Doing time at the Centre

The Rescue Centre often cared for pets while their owners were serving short prison sentences. On release, the animals were returned home. We hoped for donations from the owners, but did not always get them.

Two Collie crosses, Trixie and Leo, belonged to a man who was serving a three-month sentence in Cardiff. The police paid the dogs' boarding fees while the prisoner was on remand, but once he was sentenced the payment stopped. If a suitable kennels could not be found to care for the dogs, they would have been put to sleep - a harsh punishment to pay for their owner's crime.

In the case of Trixie and Leo, the Rescue Centre was the only charity in the South West that was able to help them. Once their owner had served his time, the police collected them and they were returned to their grateful owner.

Mean streets

Leo

Leo was an eight-year-old German Shepherd cross who lived on the streets of London. He had led a miserable life being beaten with his own lead, fed drugs to see how he would react and given alcohol to see how much it would take before he fell over. Despite all his terrible experiences, Leo was still a beautiful, loving dog and, thank goodness, his plight was brought to the attention of a caring person who could not let his situation continue. At great risk to herself, this guardian angel managed to get him away from his owner and brought him to the Rescue Centre.

We did not know how he would react to being in kennels and to living in a totally different environment. At first he was quiet and very wary of anyone going near him, but there was never any sign of aggression. He settled in quickly, but his sordid history put off many possible new owners. Eventually a lady came along with her elderly retriever and it was love at first sight. His story moved her to tears, and although nobody knew how he would cope with living in a home environment, she was prepared to take him on.

A few months later we received this letter from Leo himself:

'Well, it didn't take me long to get my four feet under the table here! Unfortunately, I cannot be top of the pack as a very fat tabby cat called Bobo holds that position. The others in the pack, Bess, Gizzy and Candy, all told me no one messes with Bobo!

My humans are fairly well trained. They take me out 2-4 times a day. I must say the food here is pretty good – varied, with chicken, meat scraps and gravy. I am ashamed to say that I have put on a little weight. I have, of course, my own car that Mum drives for me so I am not limited as to where I can go. I popped to Brixham the other day to see what it was like. Very nice!

Unfortunately, Mum can be a bit over the top. My collar matches my bed! She said I had to have a reflective collar for winter and you will never guess what colour she got – pink! I suppose she meant well.

Next door lives a lady called Granny. She has a never-ending supply of biscuits. I invite myself into her kitchen, produce a nice sit and, bingo, she gets the biscuits out. Easy really when you know how!

Finally, things are very cosy for me now and I intend to keep it that way. I am really fond of them here – even Bobo!

So best wishes and thanks to you all.

Love Leo.

PS I don't have nightmares anymore now. '

CHAPTER FOUR

New hip for a Rottweiler

Thorn was a 13-month-old Rottweiler who came to the Centre for rehoming as he was in urgent need of a double hip replacement. His family, who loved him dearly, lived in accommodation which had lots of stairs and Thorn could not cope with them in his condition.

He was a lovely, friendly boy and full of fun. It was so sad to see him try to run and play with other dogs, but because of his condition he was in a lot of pain. The operation was very expensive, with no guaranteed success, so an appeal was launched to raise the money to pay for his surgery. We circulated posters of Thorn with the slogan 'Roses are red - Thorn is so blue - He needs new hips - So, it's over to you! '

Thorn

It worked! Enough money was raised for Thorn to have his operation. He was taken to a specialist vet in Exeter for the surgery, which was successful. He

Thorn playing after his successful operation

only had one hip replaced, with plans to do the other one once he was fit. He had to be kept quiet for many weeks, with no jumping around, so a member of the Rescue Centre staff took him home to recuperate.

After six weeks, Thorn had recovered and was running and playing with other dogs at the Centre – it was wonderful to see him. Following his check up, the vet decided that the other hip didn't need to be replaced after all, so Thorn returned to the kennels and soon found a new, loving home.

Abandoned puppies

One morning we had a phone call from a local farmer's wife who had just discovered seven tiny puppies by the side of a busy road. They weren't in a box but had just been abandoned on the grass verge. I drove over to the farm, just a few minutes away, accompanied by our receptionist, Geraldine, and we were greeted by seven gorgeous identical black puppies, probably about six weeks old.

Abandoned puppies

We brought them back to the Centre, and found that although they were hungry, they seemed to be in good condition. It was a miracle that they were not hurt as the road where they had been dumped was used regularly by lorries going to the local quarry.

Puppies 1st birthday party

Then, a couple of hours later, the farmer's wife was on the phone again to say that she had just found two more pups, hiding in the hedgerow, so they joined their brothers and sisters in the special puppy kennel at the Centre. They all found new homes and one year later, we organised a first birthday party for them at the Centre. It was a joyous reunion and great fun was had by all, racing around the field. They especially loved the biscuit birthday cake that Geraldine had made for them all.

Oliver, the miracle puppy

A rather shocked gentleman arrived at the Rescue Centre one morning with a tiny bundle in his arms. He explained that several weeks earlier, when his Labrador had accidentally become pregnant, he had taken her straight to the vets to have an injection to avoid an unwanted litter. However he had just discovered two tiny puppies in her box. One had died, while the other was cold and very quiet. As the mum had shown no interest in her unexpected arrival, the man had brought the puppy to us, hoping we could help.

As he was less than a day old and had not had any of his mum's milk, the outlook did not look too good for the tiny pup. Luckily, a Collie had given birth to six puppies at the Centre a few days earlier. We decided to see if she would accept another mouth to feed, and when she was out of her kennel for some exercise, the pup was rubbed thoroughly with the scent of the other pups and snuggled amongst them.

All went well for a couple of days, but then the pup was not gaining weight and mum kept separating him from the rest of her family. So one of the girls working at the Centre became surrogate mum to little Oliver (named because he always wanted more food) and bottle-fed

Oliver 3 weeks old

him day and night. He started to gain weight and looked really healthy, but then we noticed that his eyes had a misty appearance. Oliver was checked by the vet eye specialist in Tavistock, who confirmed that he had limited vision, but we would have to wait and see as he grew how bad his sight would be. Oliver was rehomed and his new owners reported back that he was doing well.

Oliver 5 weeks old

Several years later, by coincidence, I was chatting to the girl washing my hair at the hairdressers and found that it was her parents who had adopted Oliver. He did have limited vision, but this didn't seem to bother him or interfere with his everyday life. She said that he was a very loving, special dog – and a very lucky one!

Rescued from terror

After being on the run for at least a week, a small Rottweiler bitch was brought to the Rescue Centre by the local Dog Warden. We later discovered that she had spent her life shut in a dark shed and had been used for breeding. A quivering wreck of a dog, she was terrified of everything and showed this by aggression. She growled and cowered when anyone tried to approach her and threw herself against the walls of her kennel to try and get away. It was so distressing for her and the kennel staff.

An animal behaviourist spent many hours working with her and advised that Cara, as we had named her, should have no eye contact with anyone as she took this as a threat. So staff entering her kennel had to walk in backwards. Slowly Cara's behaviour improved, but she was still very nervous and considered everybody and everything as a threat. She was brought out of her kennel every day to try to get her used to different sounds and people but she found this very stressful.

We didn't know how she would react to living in a home environment, which was probably something she had never experienced. However, Shirley, one of our volunteer receptionists, offered to take her home to see how she would cope. She already had two other Rottweilers and a cat.

Shirley wrote at the time: 'When Cara arrived, I shut the other dogs, Arnie and Tyson, in the kitchen and brought in one frightened little Rotty bitch. While I had a cuppa, Cara had a good sniff around, then I let the boys in. Although they are much bigger than her, there was no problem. They spent the next few hours getting to know each other. Cara picked her 'safety spot' in the corner of my bedroom on Arnie's bed, so I had to rearrange the beds. Not that the boys minded. Tyson, being the oldest, my big gentle giant, took on the role of nursemaid and made sure she was all right and not being bullied by Arnie. Arnie took on the role of playmate, being the big buffoon that he is.

Cara has learned more from them than anything I have taught her. She loves Bonnie, my seventeen-year-old cat. They sleep together and exchange kisses. All I wanted to do was pick her up and cuddle her – fat chance! I was her worst nightmare.

The next few weeks were difficult. I couldn't put a lead on her as she was frightened by it. Every time I went near her, she would run. Sometimes she would let me touch her for a second, then run. Talking to her didn't help either. She didn't understand words of kindness. All humans were a threat to her. What had I taken on? At least there was no malice in her.

Right, now was the time to get pushy. Leaving the boys in the hall, I went into my bedroom and shut the door behind me. Cara, in her bed, *looked at me and knew something was going to happen and quickly moved away. I trapped her in the corner and started stroking her. Her head was turned away from me but I could imagine screwed up eyes and clenched teeth. I carried on for a full five minutes, then left her alone. I would do some more later. I started a daily routine. The boys*

Cara

came into the bedroom with me and also had cuddles and Cara watched and learned. She seemed to twig after a few days. She even developed a smiley face.'

The trial run with Shirley became permanent, and after months of patience, understanding and lots of TLC, Cara gradually began to relax. Shirley told us that she could now cope with visitors as long as they were sitting down - they were no threat to her that way. She would not accept being on a lead, but would stand behind Shirley when the front door was opened. However, when Shirley moved, Cara would run to her bed, as no way was she going out into that scary world again.

Every step forward was an achievement for both Cara and Shirley, and the rest of Cara's life was spent with more love and comfort than she could ever have dreamed of when she had been shut in that dark shed.

CHAPTER FOUR

Jack, the old seadog

A very sad, old scruffy dog was taken to a local vet after being found wandering around the streets. The nurses named him Scrumpy Jack, and we soon discovered that he was deaf and blind. His fur was matted, but he was very gentle and soon won the hearts of all at the surgery.

Scrumpy Jack goes aboard	Scrumpy Jack in his life jacket

His owner could not be traced, so after a couple of weeks, the vets contacted the Rescue Centre to see if we knew of anyone who would be willing to take on such a 'special' dog. We arranged for Jack to be temporarily fostered in a home and friends of the foster carers met Jack and fell in love! His new owners, who lived on a boat for part of the year, kept in regular contact with us to let us know of Jack's progress. They wrote:

'We managed to find a really nice lifejacket on the internet with padding underneath ready for him to be lifted from our dinghy on to the back deck of our boat. We also found a hoist to mount on the deck, so everything's sorted now in our minds. He didn't worry about the boat rocking and had walks on the beach every day. He settled into a routine and felt quite at home.

Lots of people stopped to admire him and thought he was a wonderful dog. He often used to sleep in the dinghy on the beach, if it wasn't a dog friendly beach, with a sun umbrella over him to keep him cool. He went for a boat ride to Burgh Island, landed by dinghy then went for a paddle.

His cataracts look smaller again and he loved watching the sun over the sea. He seems to look around so much more now that we are sure he can see a bit better.

Our lives have changed so much for the better and we are besotted with Jack. He met another blind/deaf dog on a walk in our road so has a girlfriend now! He is an example to all us human beings. We sure do have more to learn from animals than they do from us' '

Saving a Poodle's sight

Sammy the blind Poodle had been inseparable from his owner for twelve years, so when the owner died suddenly, his future was in jeopardy. A kind friend offered temporary shelter to the bereaved little dog, but attempts to find a permanent home were soon exhausted. Luckily someone remembered hearing that Animals in Distress would help in situations

Sammy the blind Poodle

such as Sammy's, and the Rescue Centre was contacted. We immediately organised a foster home, as Sammy could not cope with being in a kennel.

Our vet examined Sammy's eyes and recommended that he should be seen by the eye specialist in Tavistock, who might be able to remove his cataracts. The first appointment was made, and the wonderful news was that despite Sammy's age, the cataracts could be removed. Although he would never have 100% vision, it would hopefully make his life easier.

Following the operation it was not clear whether Sammy could see or not, as there was a lot of swelling around his eyes and he seemed to find it easier to keep them closed most of the time. Following several weeks of medication and loads of TLC, Sammy's eyesight gradually improved and he even started chasing the neighbours' cats!

Sammy spent the rest of his life living happily with his foster mum, and the cost of his treatment was covered by a bequest from his owner's will - a final act of kindness to a loyal companion.

Barney's story

Barney

Over the years, many Greyhounds came into the Rescue Centre for rehoming, usually because of their age or because they had sustained an injury while racing. Once their racing days were over, the outlook for the future was very bleak. Greyhounds make wonderful pets as they are the most laid-back breed, and despite popular belief, they do not need a lot of exercise. Most would prefer to be lying on a couch than going out for a walk!

When Barney arrived at the Rescue Centre for rehoming, he was so nervous that he never came out of his kennel into his run unless he was on a lead, and even then very reluctantly. He did not greet visitors but sat shaking in his bed, very miserable and depressed.

Two regular volunteer dog walkers, Ron and Ann, spent many weeks coaxing and encouraging Barney out of his kennel. He was terrified of everything, and one can only guess as to the treatment he had received before he came to the Centre. Eventually, Ron and Ann built up a good relationship with Barney, and although he was still very nervous and especially terrified of children, he would go for walks with them and gradually seemed to relax a little.

After a lot of thought and heart searching, Ron and Ann decided to adopt Barney. He had probably spent all his life living in kennels, so it was a brave decision to take him and all his problems on. He was still very nervous but devoted to his new 'mum and dad'. The three of them still continued to visit the Centre regularly to take other less fortunate dogs out for walks around the country lanes. At home Barney shared his bed with a cat and became a much more relaxed and extremely happy dog.

If only there were more people like Ron and Ann around who are willing to take on an animal, despite all their problems, and just shower them with love. Barney was a credit to them.

Tearful reunion for Robbie

Robbie, a nine-year-old terrier cross, was cared for at the Rescue Centre when his elderly owner was rushed into hospital. After several weeks, it was decided by the nursing staff that his owner would not be well enough to return to her home and Robbie was signed over to the charity for adoption.

Robbie had never been used to children, so a home was found for the little chap with a kind couple without young children, although they did have grandchildren who visited occasionally. Sadly, Robbie was returned to the Centre a few weeks later for nipping one of the grandchildren.

Then we had a phone call from a nurse at a residential home in Torquay who told us that Robbie's original owner now resided there and dogs were allowed in the home. She was almost afraid to ask if we still had Robbie at the Centre, and was delighted to hear that he was still with us. A happy, tearful reunion took place at the residential home that same day.

Nine in a night

Carly, a three year old crossbreed, was heavily pregnant when she was brought to the Rescue Centre by her owners, who were moving and couldn't take her with them. A sweet little dog, Carly settled into her cosy kennel to await the birth of her pups.

One afternoon a couple of weeks later, she started to show signs of their imminent arrival. By five o'clock the first pup had arrived. My grandchildren had come round for tea that evening and witnessed the birth of two of the pups. It was an amazing experience and one they will never forget.

Staff took it in turns to keep a vigil in Carly's kennel, just in case she had any problems, but by midnight she had given birth to eight lovely pups. Everyone was ready to go home after a very long day and night, when the Kennel Supervisor, Vicki, decided to check Carly was okay for

Kennel Supervisor, Vicki, checking on Carly and her pups

a final time. Thank goodness she did. Carly had just given birth to puppy number nine, who was weak and not feeding. Carly showed no interest in the final pup, as she was exhausted by now, so Vicki took it home and syringe-fed it overnight, saving its little life. The puppy was reunited with his mum and siblings the next morning and started to feed with the others. Carly proved to be a brilliant mum, and eventually she and all her beautiful pups were found permanent, loving homes. A hard day's night but definitely worth it!

Miracle Mickey

Mickey, a 14-year-old Border Collie, arrived at the Rescue Centre in a very poor state. Since his owner had died a few weeks previously, he had been boarded in kennels while his future was decided. Animals in Distress was contacted and we agreed to take Mickey and try and find him a suitable foster home to spend his twilight years.

Unfortunately, Mickey had picked up a chest infection. Added to the stress he was

Miracle Mickey with Tracy

suffering, his health took a turn for the worse, and a few hours after he had arrived, I called the vet. Mickey lay motionless on the duvet in his kennel and at times we thought we had lost him. The vet administered antibiotics and painkilling medication and the kennel staff kept a constant vigil by his side for the next few hours, but he wasn't expected to pull through.

That same evening, Tracy, a regular volunteer and Reiki healer, brought a stray dog to the Centre that she had found wandering in the lanes. I told her about Mickey's plight, so she sat on the floor of his kennel gently touching, stroking and talking to him. Suddenly he lifted his head up and looked at her. That was when he turned the corner. The next morning, I couldn't believe the difference in him. He even came out of his kennel for a gentle walk on the grass.

A foster home was organised and Mickey, still coughing, went to recover in the comfort of a home environment. With the love and care of his foster carers, Mickey slowly recovered. Tracy followed his progress, and much to our delight, eventually adopted him. Mickey went to work with her every day in her shop in Totnes. Tracy often brought Mickey back to visit us at the Rescue Centre and his transformation was amazing – in fact, a miracle!

CHAPTER FIVE

FUR, FEATHERS AND SCALES

In addition to dogs, creatures of all kinds have been brought into the Rescue Centre over the years, from cats to cockatoos. Here are some of their stories.

A *pair of goats*

In the early days, the Rescue Centre provided sanctuary to a small number of animals which, for various reasons, could not be rehomed. Apart from providing a safe, secure home for them, we hoped that the ponies, sheep, goats and ducks would also be an attraction and encourage visitors.

I set up a sponsorship scheme to help cover the cost of caring for

Two week old Harry

these animals, and our goats, Harry and Alfie, were among the first to be sponsored. The two tiny kids arrived in the back of our Bedford Rascal van, cuddled up to our animal welfare assistant, Lisa. They were unwanted stock from a local goat farm, and as they

79

were just two weeks old, they had to be bottle fed. To have a warm baby goat snuggled up on your lap, sucking noisily at a bottle of milk, is a delightful, unforgettable experience.

Harry and Alfie soon grew into two lovable characters and were very popular with visitors to the Centre. New volunteers were 'tested out' by the pair, who would playfully butt legs or any other available body part to see the reaction they got. They soon knew who they respected, but to their great amusement, also who they could chase across the field.

Originally, they were housed in an area with a few chickens and Skippy, the rescued sheep. One young volunteer was given the task of cleaning their shed, but ended up somehow being trapped inside the chicken house, on her hands and knees, with an 'innocent'

Harry and Alfie

Grandchildren Scott and Emily feeding Harry and Alfie

Stable Animals

Harry guarding the door. I'm sure he was grinning. After her cries for help were heard, a very embarrassed youngster was released and, not surprisingly, asked if she could help clean the cattery in the future.

When the stable block was built in the bottom corner of the field, Harry and Alfie moved in and shared their new luxurious accommodation with our two rescued Shetland ponies, Jessica and Bracken, who were also part of the sponsorship scheme. They all had the use of the field for most of the day until it was time for the dogs to have their afternoon run around, then the stable animals would return to the safety of their paddock.

Oh dear, how embarrassing

One day, there were builders on site and a five-foot deep trench had been dug along the top of the field ready for new water pipes to be laid. Alfie, being the curious type, went to examine the work being carried out. Suddenly he slipped into the ditch. All that could be seen was his head sticking out at ground level.

Luckily, he was uninjured and stayed very calm as lengthy discussions took place as how to get him out. He fitted so snugly into the hole that he could not turn round and nothing could be put under him to try and winch him out.

Then the workmen had the idea of using their digger to carefully replace the soil to form a ramp at one end of the ditch. Eventually, after a lot of coaxing, a very embarrassed but dignified Alfie strolled up the ramp and carried on munching the grass in the field, as if nothing had happened.

A *hundred and forty pounds of lard*

Ho Chi Minh sunbathing

The Rescue Centre was home to two Vietnamese Pot Bellied pigs, Ho Chi Minh and later Mai Linh, who were both also part of the sponsorship scheme. At that time it was considered fashionable by some people to have a cute little piglet in their home. Unfortunately, no thought was given to the consequences when the cute little piggy grew into a huge 10 stone lump of lard!

The pigs were real characters, and very friendly. In fact Mai Linh would become too friendly when, due to hormonal changes twice a

Beautiful Mai Linh

year, she would feel the urge to mate with anything (or anybody) that moved, including the wheelbarrow used for cleaning her out!

Lisa looked after the stable animals and one of her duties during the summertime was to rub suntan cream into the pigs' skin to prevent them burning while they were sunbathing.

An old folks' home for cats

It's surprising how sometimes inspirational ideas come from the most unlikely situations. A deaf white Boxer called Rosie was brought in to the Rescue Centre for rehoming, as her owner couldn't cope with her disability. I remembered reading about a lady, Joy, who ran an animal rescue centre near Bristol and also specialised in training deaf dogs to understand doggy sign language. I contacted Joy and she said she would be happy to help, so Rosie and I made the trip to Bristol, where we received a warm welcome.

Following an assessment by Joy and her team, Rosie was accepted to begin the training that would enhance her life. When a dog was considered competent in understanding sign language, Joy found a suitable home in the Bristol area and the new owners were given lifelong support. It sounded ideal, so I signed Rosie over to Joy.

Before I made the long journey back to Devon, I enjoyed a welcome cup of tea and some homemade chocolate cake and had a wander round the site. I came across a wooden outbuilding, with pretty curtains at the windows, in the middle of a large, fenced enclosure. Outside there were shrubs, grassy areas and lots of plastic beds scattered around, but when I entered, I was amazed by the sight that greeted me. There were soft chairs, comfy sofas, rugs and cushions all over the floor, pictures on the walls and, on every available surface, snoozing cats! There must have been at least thirty moggies living together in this warm, cosy environment.

Oldies relaxing in the Greenslade Cattery

Joy explained that these were the 'oldies', mostly at least 14 years old, and therefore were difficult to rehome. A secure, comfortable home environment had been created for them and they could safely wander in and out as they pleased, depending on the weather. It was such a tranquil setting for these pussy pensioners to spend their twilight years that I was inspired to set up a similar setting in our own Rescue Centre.

Keith, 22 years old

So the Greenslade Cattery was created, based on the facilities I had seen in Bristol. Over the years, many elderly cats were accommodated in these warm, comfortable surroundings, quietly watching the world go by as they waited patiently for their forever home to come along.

Many volunteers, known as our 'cat cuddlers', would spend countless happy hours relaxing in one of the comfy armchairs in the cattery, with the cats queuing up for a warm lap to sit on. Many special needs children and adults, with their carers, would regularly

Kiki and friend listening to the radio

help to clean the cattery and enjoyed the peaceful environment.

Over the years, Greenslade Cattery became a permanent home to many pussy OAPs, the oldest resident being Keith, a scruffy 22-year-old moggy who had come to the Centre after his owner died. Another popular resident was Kiki, a Siamese cross, who came in as a stray. She had suffered spinal damage, probably following a road accident, and was paralysed at her back end. This meant she was totally incontinent and, only having the use of her front legs, hopped around like a rabbit. The Centre became her home and, being so pretty and friendly, she was a firm favourite with everyone, always being the first to greet visitors at the cattery door.

The flying goldfish

Have you ever heard of flying fish in Torquay? A lady was looking out of her window, across the garden, when suddenly a large goldfish landed from the sky and bounced across her lawn! She ran out, put the fish in a bucket of water and telephoned the RSPCA. Amazingly, the fish seemed to be uninjured following his freefall, apart from a few missing scales.

Mike, the RSPCA Inspector, didn't know where to take it, so, following a phone call, he brought it to the Rescue Centre to join the other fish in our large pond. It had probably been taken from a garden pond by a heron or seagull to make a tasty snack, but was dropped before it could be consumed. One very lucky fish!

CHAPTER FIVE

The kitten that came in from the cold

During a cold spell one December, a lady came in to Reception and said she had just seen a kitten running along the lane leading to the Rescue Centre. Some of the staff immediately investigated, but no sign could be found of the kitten. However, a cardboard box was found, hidden away under a hedge in a field, with a hole where the kitten had escaped. It was dark by 4 o'clock, freezing cold and raining but still no sign of the kitten.

Would it be possible for a kitten to survive in these conditions? Well, thank goodness, the answer was yes! Two days later the kitten was found hiding among some rubbish on a nearby industrial estate at the end of the lane. She was sneezing, but otherwise seemed okay. She was appropriately named Ice, and we thought that was the end of the story.

Lisa with abandoned kitten, Ice

However, a week later, a passer-by came into the Rescue Centre to report a sighting of a cat in a nearby wood yard, which was closed for Christmas. We obtained the keys from the owner of the yard and the frightened cat was captured the next morning. She looked very similar to Ice, although she was an adult, so we assumed that they were related and that there had been two cats abandoned in the cardboard box.

The box had been hidden in the field, so if the cats had not managed to escape, they would not have been found and would not have survived the cold, wet conditions. What a heartless act. They were lovely, friendly cats and probably mother and daughter. They eventually went to a secure permanent home together.

A cat behind bars

When builders were working on an extension at nearby Channings Wood Prison they noticed a cat hanging around, but they couldn't get near to it. They assumed it was a feral cat, but as the work progressed it stayed on the building site. By the time the work was finished, the poor cat was trapped firmly behind bars, with no escape.

A prison officer rang the Rescue Centre and asked if they could borrow our cat trap to catch it. A cat trap is a cage with a trap door which closes behind the cat when it goes in for food and does not harm the cat at all. Within a couple of days, Channey (as we called him) was brought to the Centre in the trap. When released into the cattery he was found to be a very affectionate, if shy, moggy. He was neutered and vaccinated and adopted by one of the prison officers and his wife. I love happy endings!

A bin full of rats

One of the rescued rats

One day, a lady arrived at the Centre with a plastic bin in the back of her car. Quite innocent, one would think, until the lid was removed to disclose 14 tame rats of all colour and sizes! She had seen two lads releasing them on waste land, so she promptly gathered them all up, put them in the bin and brought them to us.

87

Thankfully, they were rehomed quickly to a 'ratty' lady who lived in Paignton and had helped us out previously. The lady drove up to the Centre with one of her rats sitting on her shoulder!

One good quack deserves another

The first duck to arrive at the Rescue Centre was Ernie, a small white call duck. He was found wandering along a village street, but was obviously used to being handled as he was quite happy to be

Ernie in his de luxe' pond

picked up and brought to the Centre. Suitable accommodation was quickly organized and Ernie was soon splashing around happily in his own 'pond', a child's paddling pool.

Ernie was soon joined by other needy ducks, many of whom were brought in following the loss of their partners, as their owners were concerned that they were pining and lonely. A duck enclosure was created next to the chicken area, with a large pond and plenty of room for the ducks to waddle around.

A local couple were on their way to North Devon when they came across a Muscovy duck wandering in the road. She seemed unharmed and very friendly, but her home could not be found. They popped her on the back seat of their car and, later that day, arrived at the Centre. She seemed to be unflustered following her journey and settled in quickly with our two other Muscovy ducks, Howard and Hilda.

Howard and Hilda

Two mallard ducks were brought to the Centre for their own safety, by a member of the public, as they had identical twisted wings, which stuck out an angle and prevented them from flying. Staff named them 'The Twisted Sisters'. Although they were wild ducks, they soon adapted to life in their new safe home.

One day, there was great excitement when one of the Muscovy ducks was found to be sitting on eggs. A special enclosure was set up for her so that the ducklings would be protected from birds of prey, etc. and we

The duck area

waited patiently for the big event, the first ducklings to be born at the Centre.

Surprise surprise - when the eggs finally hatched, three beautiful baby chicks emerged. One of our chickens had obviously decided that the duck enclosure would be a safe place to lay her eggs, and the Muscovy duck kindly obliged and became foster mum. Three very confused chicks!

'Lucky day '

It must have been Gordon the Gerbil's lucky day when he was discovered on a garden patio by Timmy the cat, who took him carefully indoors and proudly presented him to his owner. Gordon was brought, unharmed, to the Centre, and as he was not claimed, soon found a new home. Fortunately for Gordon, Timmy must have had a big breakfast that morning!

Rabbit in a box

When a lady in Dawlish put her rubbish out one morning, she noticed that a cardboard box beside the dustbins was moving. When she opened the box, she discovered a pathetic little rabbit. He was in a terrible condition and she took him straight to the local vet. He needed treatment for his painful, infected eyes and his teeth were also in a poor condition. The cost of the veterinary treatment was high and, although the lady

Abandoned Taffy

would have liked to have helped him, she just couldn't afford to and,

Taffy with his wife, Debbie

besides, she had no garden to keep him in.

So she brought the little rabbit to the Rescue Centre to be cared for. He was named Taffy and treatment began immediately to make him feel more comfortable. His poor eyes were so mucky he could barely see, so they were gently bathed and he was booked into the vets to have a dental.

Taffy soon recovered and was a friendly little bunny. He certainly didn't deserve the neglect he had suffered. When he was fully fit (and neutered!) Taffy was introduced to a pretty little rabbit called Debbie. They fell in love and the pair were rehomed together to live happily ever after!

Cruelly abandoned

Over the years, many animals were abandoned at the gates of the Rescue Centre, usually out of hours. Sometimes boxes were left in the car park containing either kittens or rabbits and even once, a cockatiel and two budgies.

One cold, dark December morning when hubby Dave was leaving for work at around 5.00 am, he discovered a Staffordshire Terrier tied to the gatepost. The sad little dog was soaking wet as it had been raining heavily. He had probably sat there quietly all night waiting for his owner to come back. How utterly heartless.

On another occasion when Dave was off to work, he nearly tripped over an old hessian sack tied up with string that had been left outside our front door overnight. On closer inspection, whatever was inside seemed to be writhing around – oh no, surely not a snake? When the sack was very gingerly opened, four little faces peered out – kittens! What a relief! Despite being very frightened and hungry, they all survived their ordeal and were later rehomed.

Hungry Kitten

Early one morning, a cardboard box was left at the gates of the Rescue Centre. Inside were two young chickens, frightened but healthy. On the box was written, 'Please look

Richard and Judy (Julian)

after us. We are three months old and our owner cannot keep us'. We named them Richard and Judy and they settled into the paddock with the other resident chickens and ducks. However, as time went on they had to be renamed Richard and Julian, as they both grew into beautiful cockerels and crowed in unison every morning!

Garfield, the skydiving kitten

A young man telephoned the Centre one morning and explained that he had recently bought a kitten from a man in a pub. The

Kitten with broken leg

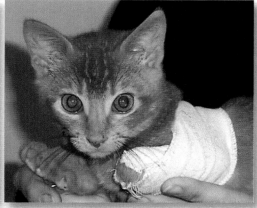

Garfield, the kitten with a broken leg

previous evening the kitten had jumped off the top of the wardrobe and now it was limping and crying. He couldn't afford to take it to a vet so he asked if we could take it from him.

When the pretty little ginger kitten arrived at the Centre, it was obviously in a lot of pain so was taken straight to see our veterinary surgeon. Following an x-ray, a broken leg was diagnosed and the poor little mite, whom we named Garfield, had to have his leg strapped to his body to stop any movement, as it was too small to be put in plaster.

Garfield was very brave and spent the next six weeks hopping around the cattery on three legs. He eventually made a complete recovery and found a lovely new home.

Dumped in the woods

A lady was walking with her dog in Stover Park when she came across a guinea pig, hiding in the undergrowth. She managed to capture it, put it in the boot of her car

Guinea pig with poorly foot

and bring it straight to the Rescue Centre. On examination, the frightened little chap had a nasty abscess on his front foot. Following a visit to the vet, he was put on to antibiotics and the dressing on his little paw had to be changed daily. Following lots of TLC, he made a complete recovery. We guessed that his owners must have dumped him in the woods as they couldn't afford the fees or couldn't be bothered to take him to see a vet for treatment.

A vanload of guinea pigs

It had been a quiet autumn day at the Rescue Centre and the staff were clearing up and getting ready to go home when suddenly a police riot van pulled into the car park. Mutiny in the village? Surely not! Two burly policemen got out of the van and walked into reception, where several of us were waiting, wide-eyed with anticipation.

They explained that earlier that day a gentleman had been detained for a considerable length of time in prison. He kept guinea pigs, and had asked that they could be found new homes. We agreed to take them in for rehoming and followed the policemen round to the back of the van, where they showed us two large cardboard boxes. When we opened them, the sight was incredible. At least 40 guinea pigs of various colours and sizes were squirming around, piled on top of each other! Males and females were all in together, so there were many family groups but who belonged to who was anyone's guess.

So the mammoth task began of sorting them out, firstly into boys and girls and then to identify those which looked as if they were pregnant. Some were tiny and obviously just a few days old. Luckily guinea pigs are born quite independent and can feed themselves from an early age, so the females that didn't look pregnant were put in an area together with the young ones. They seemed to sort themselves out and all the babies survived.

During the following weeks, several of the females gave birth and the total count was well over fifty! Luckily, guinea pigs are a popular pet and all were eventually rehomed.

The cat that feared humans

Winston, a small black cat, was brought to the Rescue Centre as a stray. He was terrified at being in the cattery as he was thought to be part feral (ie, wild) and was therefore very stressed at being confined. The decision was made to release him and we hoped that he would make his home at the Centre.

For several years Winston hid from us humans during the day but came into the reception area at night to feed and sleep. His little black face was often seen peering from under hedges, and eventually he befriended one of the Centre's cats, Tich, and started to follow him around. Sadly, Tich was killed by a car in the lane leading to the Centre, so Winston transferred his affections to Bags, another

Sky, Stranger and Winston

kindly, elderly Centre cat. Bags eventually died too, of old age, and once again Winston had to find a new friend. During all this time he would not let a human within six feet, scooting off to safety if anyone came too close.

Winston, now elderly himself, became inseparable from his new companion, a big black cat called Stranger, whose story appears below. Stranger often slept in a basket in my office and Winston decided that this was a good idea now he himself was in his twilight years. It had taken about ten years for him to trust anyone not to touch him, so he moved, very warily, into my office with Stranger.

When Stranger's health started to fail due to old age, they both decided to make their home in our bungalow on the site, completely their own decision. They took up residence under the kitchen table and snuggled up together in a cosy bed.

Tia with Winston

When Stranger died, Winston decided that he would stay in his comfortable environment on his own. Gradually, he would allow us to stroke him very gently, around his ears. After a few months all his fears seemed to disappear and he finally relaxed in his own safe home. He became quite a character in his twilight years and very demanding when it was his dinner time!

It had taken a very long time, but Winston had finally experienced the feeling of being loved and cared for.

A stranger comes to stay

Not long after the Rescue Centre opened, a large black cat limped up the drive with blood dripping from its paws. His torn pads were stitched and he was pampered in the cattery until his health was restored. Stranger (so named because we never found out where he came from) decided to stay.

Over the years, Stranger became a favourite with the staff and visitors and was often seen being cuddled by an unsuspecting admirer, dribbling on their shoulder and down their back. He would leap up into your arms for attention, even when you weren't looking. His friendly, laid-back approach meant he soon used up his nine lives, plus many more, over the years. Stranger trusted everyone and had no fear, which meant he was chased on several occasions by not-so-friendly dogs.

Stranger was the subject of another article I wrote for the charity's magazine, entitled A Day in the Life of Stranger, in his own words:

'My day usually begins when Keith (kennel worker) arrives at the Centre at the crack of dawn, turns the lights on and the dogs start barking with excitement knowing it is nearly breakfast time. After a good, long stretch, I follow him to the kitchen and make a general nuisance of myself until he eventually opens a special carton of food for me.

I have a friend called Winston. He is a little black chap and although he loves to be with other cats, he doesn't trust the humans at all. I have tried to find out why, but he just won't talk about it – he must have had a pretty bad experience before he came to the Centre. A long time ago, my old mate Bags (another resident) took little Winston under his wing (or should it be paw?) and they were inseparable. Sadly Bags has now gone to the 'big cattery in the sky' and Winston now follows me everywhere and we sleep curled up together in a cosy basket in the office.

However, he won't come with me into one of my favourite places – the rabbit enclosure. I like to supervise the cleaning of the hutches in the morning and I have always got on well with the bunnies and guinea pigs. Unlike Winston, they have lots of stories to tell me – usually very sad tales of being kept on their own in small, miserable hutches with nothing to do all day. Quite a few of them arrive with bad attitudes towards the humans. After a few days, they realise that there is more to life than sitting gazing out through the wire of their hutches and that in fact there are some humans who actually love them, talk to them, brush their fur and make them feel very special. It brings a lump to my throat when they eventually go off to their new homes, but I know they will now be cared for, for the rest of their lives.

I had a most embarrassing incident the other day. I was in the rabbit enclosure one morning, supervising as usual. Sometimes the humans let the rabbits run around on the floor while their hutches are being cleaned out, so I just went to have a chat and suddenly this particular rabbit attacked me and had me by the throat! After I was rescued I just couldn't look the humans in the face and made a hasty retreat to regain my composure. Thank goodness none of the other cats knew what had happened – it would not be good for my big cat image!

I have a job at the Centre which nobody really likes to talk about, so I just get on and do it quietly – and I'm pretty damn good at it too! My official title is Chief Mouse Catcher, and although the humans know it has to be done, they prefer not to watch, so I try to be discreet so as not to upset them.

Stranger and Winston

Tragically, one day, while he was asleep in the sun, Stranger was attacked by a dog being taken out for a walk and one of his front legs was broken in several places. Because of his age, his brittle bones would not mend, so the leg had to be amputated. He then decided he needed

Sky, Stranger & Winston

more comfort in his old age, so he moved into our bungalow, bringing with him his best pal, Winston. Stranger's eyesight was failing, but he still managed to get around on his three legs.

They took up residence in a basket under the kitchen table, but gradually moved into the lounge and snuggled up on the sofa together. Sky and Tia (our dogs) treated Stranger and Winston with respect, following a few taps on the nose from Winston, and they all lived happily together.

Snatched from a watery death

One day we received a distress call from a young man whose job was artificial insemination on farms. In the course of his work, he was visiting a farm and saw that they were about drown a tiny, helpless kitten. He could not just watch and let this happen so he took the kitten and called us for help.

He brought the kitten to the Rescue Centre and we were shocked at what we saw. He presented us with a little bundle wrapped in an old jumper. The kitten was

Dennis at 3 weeks old

Dennis and Tia having a chat

soaking wet with a discharge from his eyes and nose and was having difficulty in breathing. He needed round-the-clock nursing and feeding. The cattery supervisor, Lisa, took on the task of nursing the little kitten back to health, which required a lot of time, dedication and knowledge. He spent the day in my office and went home with Lisa overnight. Sometimes he came to stay at home with me, to give Lisa a break, and my grandchildren loved to help with his bottle feeding. After several weeks, Dennis, as he was now called, recovered fully and was ready for rehoming.

Then - disaster! Dennis developed ringworm, so he could not leave the cattery, as this is a very infectious condition. Eventually, six months after he arrived, Dennis had grown into a beautiful, healthy cat and was found a wonderful new home.

'Please take care of my cats'

Sugar

Three cats were dumped one night in a cardboard box on the edge of the main road near to the Rescue Centre. A note had been written on the box saying 'Please take care of my three cats. They are all friendly and I love them but can't keep them'.

Unfortunately, by the time the box was discovered the next morning by a member of staff, all three had escaped from it. A search was started and two of the cats, females, were discovered hiding in the bushes at the end of the lane. The third escapee, a white long-haired male, was seen tracking across a nearby field, but then disappeared.

Over the next few months many sightings were reported of Sugar, as we called him. He came to reception every night to eat and sleep, but was gone again by the morning. He made his home under a large tree near the Rescue Centre, but although he was friendly with the resident cats, no human could get near to him.

The months turned into years. Sugar's routine was to wait until the last member of staff had gone home, then eat, sleep and play with the rest of the resident cats quite happily until the morning, when he would take off and spend the day lying in the sun in the field next door or sheltering from the rain under his special tree.

Then a miracle happened. (We had lots of those at the Rescue Centre!) We noticed that Sugar appeared unwell and lethargic, and a red patch could be seen under his chin. It was time for action. The cat trap was set near his tree and within a few hours he was caught, unharmed. Assuming he would not want to be handled, he was very carefully released from the trap into a pen in the cattery. To our utter amazement Sugar purred and, after a few hisses, allowed himself to be stroked very gently. He had an abscess on his neck which was treated and he made a good recovery. When he was fully recovered he was released to live at the Rescue Centre alongside the other resident cats.

The bullock who played football

William was a six-month-old Jersey bullock who was born on a nearby smallholding. His owners could not afford to keep him with his mum, so he was brought to the Centre, where he shared the field with our Shetland ponies during the day and slept in the goats'

Dave and William the calf

paddock at night. He was extremely tame and loved to be tickled under his chin. His favourite pastime was playing football with Dave in the field, which he did with great skill - for a bullock!

As he grew up, it became obvious that William needed more space, but rehoming him to a farm where he could have been sent for slaughter was unthinkable. He had become a real pet, and everyone loved him.

We contacted Hillside Animal Sanctuary in Norfolk, which was founded to help all animals in need, but especially farmyard animals. They just happened to have a lonely Jersey cow, Tinkerbell, who didn't get on with the rest of the herd, so they agreed to take William to be a companion for her. It was a match made in heaven and the two of them lived happily ever after.

The cat that played with the dogs

Sky and Ginger chilling out

Ginger was a fabulous cat. He became a resident at the Rescue Centre and took on the role of Mouse Catcher in Chief with great success. He decided he would live in the bungalow with us and spent most of his days sleeping on our bed or stretched out on the sofa in the lounge with the dogs. He often visited the rabbit enclosure, where he knew he would receive special treats from Linda and her team and would always find a comfortable spot in the sun to snooze away the time.

Every evening after the Centre had closed, Ginger would follow Sky, Honey and Paxton down to the field and chase around with them, just like a dog. When I threw Sky's ball for her, Ginger would chase it, often getting there first, and Sky wouldn't attempt to

"Think I will stay indoors today!"

Ginger, snug inside

retrieve her ball until Ginger got bored and walked away.

He was also very polite, as when he received his bowl of food or fresh water, he would meow 'thank you'. Yes, really! A real character.

Poultry in motion

One day a rusty old van containing over seventy chickens that had been rescued from a factory farm drove into the Centre car park. We had agreed to take some of these birds from another charity to live with our own chickens, who at that time had their own paddock, next to the rabbit enclosure.

When the back of the van was opened I wasn't prepared for the sight that greeted me. Cages full of what looked like oven-ready chickens were stacked to the roof. Some of the birds had died en route. Although we had only agreed to take a dozen or so, every single live chicken was unloaded and taken to our chicken area. It was standing room only, but at least they were out in the fresh air, and most of them started happily pecking at the soil.

They were given plenty of food and water and later that evening I went out to check that they were comfortable for the night. To my horror, half of them had disappeared! I frantically checked the fencing, but that was all sound, so where had they gone? Then I looked up and there they were, snuggling up together, roosting high in the branches of the trees around the paddock. I honestly didn't know that chickens could fly!

The next morning, after they had all returned to earth, the escapees had their wings clipped for safety and adequate perches were put around their enclosure. Once they had all regained their health, most were rehomed. They produced masses of eggs for their new owners.

'I'm a sheep? No kidding!'

Skippy

On a local farm, an orphaned sheep had been bottle-fed by the farmer's wife. When he was old enough to rejoin the flock, he didn't identify himself as a sheep and would not settle with the other lambs. The farmer told his wife that the lamb would have to go to market, but being a kindly soul, she couldn't bear the thought of that, so, unknown to her husband, she secretly brought him out to the Rescue Centre. Skippy was very friendly and spent the rest of his days living happily and safely with Harry and Alfie, the goats.

Lamb of God

A family from Newton Abbot went to Cornwall for a holiday on a farm. While they were there it was lambing time, and a tiny lamb was born to a mum who rejected it. As she was so small, the farmer didn't want to bottle-feed her, so the family brought the lamb home with them.

Annie lived quite happily in the family's semi-detached house in the middle of town with two cats, one dog and four boisterous children. The parents realised that it was not the ideal situation in which to raise a sheep, so they contacted the Rescue Centre. When I arrived in our van to pick her up, the little lamb was peacefully asleep in the dog's basket in the kitchen!

Skippy and Annie

Annie found it hard to adapt to living with our goats, Harry and Alfie, and Skippy, our male sheep in their paddock, because she had become so domesticated. However, after a couple of weeks, she settled down and was a very popular little lamb, especially when it was feeding time and our young volunteers loved to take it in turns to bottle feed her.

Annie walked quite happily on a lead and accompanied me on several occasions to fundraising events. At Christmas, she even took part in a carol concert in the church at Marldon village.

A new home for Barbara and Hyacinth

One day a member of staff noticed a cat carrier had been left outside one of the catteries at the Rescue Centre. It contained two very large rabbits. A note was attached which said 'Please look after Barbara and Hyacinth. They are mother and daughter and we love

Barbara & Hyacinth

them very much'. Both the rabbits were in good health and were well fed. They were very close, as rabbits usually are, and snuggled up together when put into a hutch.

Following some publicity in the local newspaper, a new home was found for this lovable pair of bunnies and they spent the rest of their life together in a secure, loving home.

A kitten comes back from the brink

When a pretty little stray cat was brought into the Rescue Centre, it soon became obvious that she was expecting kittens. She was named Rosie and soon started to show signs of her kittens' imminent arrival. She gave birth overnight to three kittens, only one of which survived.

Rosie was very distressed and showed no interest in her tiny surviving offspring and, in fact, kept pushing him out of her bed on to the cold floor. Numerous attempts were made to get her to feed him, but he was getting weaker and colder and I thought he wasn't going to survive. I took his limp little body indoors to try and warm him up and get some milk into him via a syringe. He was fading fast, so I wrapped him in cotton wool and put him into our airing cupboard, which was the warmest spot I could find. I thought at least he could pass away in warmth and comfort. I checked him several times and his breathing became very shallow.

Then a miracle happened (again)! I quietly opened the airing cupboard door to see if he was still alive and there he was, sitting up amid the cotton wool, looking at me! I couldn't believe it. I quickly fed him some warm milk, which he took well. He gradually regained his strength over the following weeks and turned into the most stunningly beautiful kitten I have ever seen. I called him Boomer, short for boomerang, as he had come back from the dead!

Basil and Boomer

I would love to have kept him myself to live at the Rescue Centre, but at that time not all the dogs we had living indoors were cat friendly. So one of our

Beautiful Boomer in Helen's garden

volunteer receptionists, Helen, adopted him and he lived happily with her and her three other rescued cats.

Hello Mr Chips

Mr Chips, an elderly tabby cat, was brought to the Rescue Centre for rehoming by a distressed gentleman following the death of his wife. Mr Chips had been her cat and had spent most of his life in her company, being loved and spoilt. After she died, Mr Chips stopped eating and lay next to a photo of her on the dresser every day. He was pining away.

When he came into our cattery, he was very quiet and depressed and spent most of the day hiding behind his bed. I contacted the

Lisa and Mr Chips

local newspaper, which agreed to feature his story. A local couple read the article and came to the Centre the next day to meet Mr Chips, who came out from behind his bed purring and took to them immediately.

Hasty arrangements were made for his adoption and after a few days, the couple reported that Mr Chips had settled in well and started eating and enjoying life again.

Rescued from the flood

Early one December morning we had a call on the emergency phone from Teignbridge Council, asking if we could help following a serious flooding of a mobile home site in Dawlish Warren. Residents had been evacuated overnight to the local leisure centre, along with their pets, which included dogs, cats and a cockatiel. They had just had time to grab essential items, and therefore needed blankets, bowls, litter trays and food for their pets.

When I arrived everything seemed to be well organised, with the Red Cross dishing out hot food and drinks to the evacuees. There was a real 'Dunkirk spirit' at the leisure centre and the animals were very calm, although a little confused.

One elderly gentleman was very concerned, as his cat was still in his home. He had been rescued by helicopter when the floodwaters were lapping at his front door. The local dog warden and I went back to the site with the gentleman and the scene was amazing. It looked like a film set.

The coastguards took the man in a dinghy to his home, where his frightened cat was hiding under a settee. He was safely rescued unharmed, despite his ordeal. Another pet owner had to return by dinghy to collect his dog's essential medication which had been left behind.

It was a long, tiring but very rewarding day with no casualties – human or animal!

THE END